PENGUIN BOOKS

THE RIGHT LIFE

Remo Largo is a Swiss paediatrician and educationalist, internationally known for his long-term studies of child development. He has written several popular books on the topic. He lives in the Swiss village of Uetliburg with his three daughters.

T0200151

REMO H. LARGO

The Right Life

*Human Individuality and its Role in Our
Development, Health and Happiness*

PENGUIN BOOKS

PENGUIN BOOKS

UK | USA | Canada | Ireland | Australia
India | New Zealand | South Africa

Penguin Books is part of the Penguin Random House group of companies
whose addresses can be found at global.penguinrandomhouse.com

First published in German under the title *Das Passende Leben* by
Fischer Verlag GmbH, 2017
This translation first published by Allen Lane 2019
Published in Penguin Books 2020
002

Typeset by Jouve (UK), Milton Keynes
Printed and bound in Great Britain by Clays Ltd, Elcograf S.p.A.

A CIP catalogue record for this book is available from the British Library

ISBN: 978-0-141-98533-6

www.greenpenguin.co.uk

MIX
Paper | Supporting
responsible forestry
FSC
www.fsc.org   FSC® C018179

Penguin Random House is committed to a
sustainable future for our business, our readers
and our planet. This book is made from Forest
Stewardship Council® certified paper.

For
Eva, Johanna and Kathrin
Jana and Remo
Aroǹ and Miguel
Brigitt

We are not the playthings of a blind external power, but the sum of the gifts, weaknesses and the other things we inherit, which a person brings with him. The goal of a meaningful life is to heed the call of this inner voice and to follow it as far as possible. The path would thus be: recognize yourself, but do not judge or desire to change yourself. Rather, let your life most closely approximate that predetermined shape, an inkling of which you already carry within you.

Hermann Hesse (1877–1962)[1]

Contents

Introduction: Living Together as Individuals   xiii

1 Human Biological and Socio-cultural Development   1

    All living things developed out of one another   3

    Adaptation and change is a principle of life   10

    Why human evolution has taken such an unusual course   22

    Socio-cultural evolution   30

    Fundamentals for the Fit Principle   41

2 The Combined Effect of Genetic Predisposition
and Environment   47

    How genotype and environment work together   48

    Children help to plot their own course   62

    Fundamentals for the Fit Principle   70

3 Developing into Individuals   77

    Our brains   78

    What maturity and experience contribute to brain
    development   83

    Genuine curiosity and learning that lasts   90

    Fundamentals for the Fit Principle   102

4 Basic Needs that Shape Our Lives   112

    Our basic needs   114

    Physical integrity   116

    Emotional security and affection   118

CONTENTS

Social recognition and social status   121

Self-development   122

Striving for achievement   124

Existential security   126

A unique profile of basic needs   129

Fundamentals for the Fit Principle   132

5 Competencies We Want to Develop   136

What we mean by intelligence   137

Social competencies   141

Linguistic competencies   162

Musical competencies   170

Figural and spatial competencies   171

Logical and mathematical competencies   174

Temporal and planning competencies   177

Motor and kinaesthetic competencies   180

Physical competencies   181

The unique interplay of competencies   184

Fundamentals for the Fit Principle   187

6 Our Ideas and Beliefs   194

The essence of our ideas   195

Conscious and unconscious ideas   202

Fundamentals for the Fit Principle   210

7 From Nature to the Man-made Environment   215

Estranged from nature   216

The transformation of our social environment   222

Fundamentals for the Fit Principle   228

8 Living the Right Life: The Fit Principle   233

      How the Fit Principle came about   234

      The elements of the Fit Principle   238

      Free will and the meaning of life   256

9 Misfit Constellations   260

      Misfit constellations are as varied as humans
         themselves   261

      Managing Misfit situations   274

      Reassuring people and rethinking the environment   287

10 Changing Times   291

      General insecurity   292

      Fit and Misfit in society and the economy   295

      Family and community for a fitting life   303

      Living a fitting life is a human right   310

      Appendix: The Zurich Longitudinal studies   313

      Acknowledgements   321

      Bibliography   323

      Notes   335

Introduction:
Living Together as Individuals

*Every human being is unique. Exercising your
individuality amounts to the meaning of life.*

> *Be true to thyself now that thou hast learnt what manner of
> man thou art.*
>
> Pindar (c. 518–442 BC)[1]

I love to observe people of every age – on a summer's day in Zurich's
Old Town, for instance. The Münsterhof – Minster Square – is a
constant bustle of strolling tourists, hurrying businesspeople, locals
exchanging news and children playing. I am fascinated by the diversity in their faces and characters, the different ways that children,
adults and old people interact with each other: the differences in their
body language when the grown-ups greet one another or the little
ones chase each other around; and the differences in the adults' interest
in the venerable Fraumünster Church or the shop window displays.
I never get bored of it. I am sure that no two people with exactly the
same appearance and mannerisms will ever walk across this square,
because I know that each and every one of the nearly 8 billion people
who currently inhabit this planet is unique. This diversity is by no
means unusual. Plants and animals are just as diverse within their
own species. But what makes humans special, and what makes me an
observer of them, is that we are the only ones – thanks to our highly
developed intellectual abilities – who are aware of our own individuality and the diversity of our species.

We begin to recognize ourselves as independent beings from when
we are just two years old. Over the years that follow, we learn to
imagine and empathize with other people's emotions, thoughts and

behaviour. And in so doing we learn that every person has their own characteristics, talents and ideas. By the time we have started school, at the latest, we begin to compare ourselves with others, and we continue this behaviour all our lives. As adults we measure ourselves against other people, in terms of appearance, or professional and social standing, or performance and income. We delight in our strengths and deplore our weaknesses. We wonder how other people see us. And again and again we are thrown back on ourselves: what do we have to accept about ourselves as 'given', and what can we change if we just try a little harder? Over the years, however, we realize that there is no ideal path to follow through life – even if countless self-help books confidently promise us such a thing. And this book can't offer an 'ideal path', either. Rather, it is an attempt to give readers an understanding of human individuality and the various ways humans try to survive in this world. Because we still have difficulty with the concept of individuality. We think and behave as though we were all the same, with the same needs and the potential to achieve the same things. But this is a long way from the truth. There are no universally applicable rules for living in harmony with your environment. This is a challenge that we must all overcome in our own way.

Exercising our own individuality isn't the only challenge; we also have to deal with the diversity among our fellow humans and how different they are from us. Imagine if we were all the same: the same height and weight, the same looks, born with the same feelings and talents and the same needs. Life would be monotonous – though there are some problems that diversity creates within families, schools and societies that wouldn't exist. But without diversity *we* wouldn't exist either, and nor would any other life form. Diversity and individuality are the fundamental preconditions for all life.

Just how diverse humans are and how many difficulties this diversity brings with it are what made the most lasting impression on me in my thirty years as a researcher and practising developmental paediatrician. From 1974 to 2005, I had the privilege of continuing a large-scale research project begun in 1954 at the Zurich Children's Hospital. The Zurich Longitudinal Studies followed more than 700 normally developed children from birth to adulthood, in two consecutive generations, documenting the development of each child in areas such as motor skills and language. Our motivation for under-taking such extremely labour-intensive studies was the conviction that only when we are familiar enough with the diversity and the typical patterns within normal development can we do justice to children's individual needs and abilities, and support their development effectively in our roles as parents, therapists and teachers. In fact, when we looked at the data on the various areas of development, it revealed that there was no ability, no behaviour, and no physical or mental characteristic that is developed to exactly the same level in all children. Children differ in terms of height and weight at every age. They require differing amounts of sleep and eat different quantities of food. Some children take their first steps at ten months; others don't walk until twenty months. Occasionally children are interested in letters when they are just three or four, though most learn to read between the ages of six and eight, and some still have difficulty reading well into adulthood. In every respect, this diversity increases steadily as children get older, and it continues to increase – though to a lesser extent – when they become adults. Some adults' understanding of numbers never gets beyond primary-school level, while others have logical and mathematical abilities that enable them to perform complex IT tasks.

This means that humans have very different resources with which to overcome the challenges, large and small, that life throws at them. Take Luca, for example, who came to my clinic with his parents. He felt he was a failure because, at the age of nine, he still couldn't read. He was painfully aware that he was unable to fulfil the expectations of his parents and teacher. Luca's sense of wellbeing was significantly impaired, and he reacted by becoming distracted and fidgety. In the course of my work, I have seen thousands of children like Luca, who have been brought to us because they deviated from the 'norm'. They were suffering from a wide variety of developmental and behavioural abnormalities such as waking up at night, poor coordination or under-developed social skills. Though they often didn't say as much, their parents and teachers wanted us to help bring these children back in line with the 'norm' – which, as our many years of experience had taught us, cannot be done. For us, the children's real problem was that, because they didn't conform to ideas of normality, they weren't allowed to be 'themselves'. And so we tried to help them by understanding their individual needs and abilities. We would then discuss with parents and other care-givers how best to support each child with their individual strengths and weaknesses. It wasn't a simple undertaking; the adults had their own expectations of the child, their own ideas about his abilities and, above all, about the things he should have been able to achieve. But when we did manage to tune the adults into children's individual needs and abilities, the children's mental and physical condition improved and their willingness to learn increased.

Exercising your own individuality remains a constant challenge even in adulthood. Let's take someone who works for a bank, for example. Just like Luca, the schoolboy, her wellbeing is impaired when she can't achieve what she expects of herself in the workplace, and what her managers and colleagues demand of her. She feels overwhelmed, becomes exhausted and, in the worst-case scenario, starts to suffer from burnout. Her wellbeing won't usually be improved by trying to improve her performance, through additional training, for example – though that is the path employers often take. What she needs is for people to respect her individual talents and bring the demands of her work into line with her abilities as far as possible. The same problem of adaptation arises when people aren't being challenged enough; the sense that what they have achieved is unsatisfactory, or even meaningless, can also significantly impair their sense of wellbeing.

Several times a day, in both research and clinical practice, we were confronted with the question: why does one child feel comfortable and develop well, while another's wellbeing is affected, causing abnormal development? We almost always found the answers to this question in the degree of harmony between children and their environment. Thus, for example, we realized that sleep disorders frequently occur because parents have the wrong idea about how much sleep their child needs. At the age of twelve months some infants sleep for fourteen hours a night, while others find nine is enough. If the parents manage to tune into their child's individual sleep requirements, the sleep disorder disappears. Over the years, observations like these taught us to pay attention to whether there was harmony between the child and his environment, in all areas of development – and, if not, to recognize how the child was being affected by the disharmony and how this could be remedied.

Questions of human individuality, and of the interaction between human beings and their environment, have essentially been my pre-occupations since I was a teenager. At the age of thirteen I was confined to bed for eight weeks, and during this time I devoured Leo Tolstoy's *War and Peace* and Dostoyevsky's *Crime and Punishment*. I was so fascinated by the empathetic and lifelike portrayal of differ-ent human characters and the dramas that played out between them that – once I was well again – I read my way through all the Russian literature that was available in German. Since that time, I have never stopped thinking about why people are so different, what shapes their lives, and what ingredients go to make up a human being. I began my medical degree at the University of Zurich in 1963, hoping it would give me a more profound understanding of humanity. But my degree was a strange experience for me: I encountered an immense number of mental and physical phenomena of all kinds, but my cata-logue of questions grew longer rather than shorter, and I still hadn't gained a deeper insight into the essence of humanity. In the dec-ades that followed I went in search of a holistic view of humans, engaging with the most diverse specialisms, in particular evolution-ary biology, philosophy, pedagogy and psychology. I read the writings of geniuses like the philosopher Immanuel Kant and the evolutionary biologist Charles Darwin, the educator Maria Montessori and the psychologist Jean Piaget. But again and again I was disappointed. Their writings illuminated important partial aspects of the human condition, but I still didn't have a comprehensive view.

Over the course of forty years the things I have learned in the hos-pital and through my research, along with findings from diverse areas of study such as genetics and sociology, gradually came together – like pieces of a puzzle – to form a complete picture. I called it the Fit Principle. What it amounts to is that *every human, with their indi-vidual needs and talents, strives to live in harmony with their environment*. The Fit Principle rests on a holistic approach that sees the diversity of humans, the unique nature of each individual, and the interaction of individual and environment as the basis of human existence.

How successful are humans at exercising their individuality in harmony with their environment? More and more people are feeling overwhelmed by the struggle to live a fitting life. Children are expected to fulfil their parents' often exaggerated expectations, and schools place them under an unbearable pressure to achieve. Adults fight a constant battle to balance family life with work and the growing demands of the economy. Older people, especially those who live in care homes, suffer from a lack of emotional security and from social isolation. People of every age are feeling increasingly at odds with the world and less able to live a life that corresponds to their individual needs and talents. On an individual level, the Fit Principle can help people find their way back to their individuality. And, on a larger scale, it can help people to start reshaping their society and economy so that they can live as successful a life as possible.

This book describes a wide arc from the beginnings of evolution to modern humans, and the following brief overview of its ten parts is designed to give readers an introduction to the intrinsic connection that exists between subjects as different as evolutionary biology, nature and nurture, human development and the Fit Principle.

CHAPTER I: HUMAN BIOLOGICAL AND SOCIO-CULTURAL DEVELOPMENT

Humans are related to every living thing on this planet.

There are many aspects of our own lives that we can only understand and explain to ourselves when we call to mind what has happened to us in the past. In the same way, a glance back to the distant origins of humanity can also help us to understand our (modern) nature.

In the Old Testament, in the first Book of Moses, the creation story tells us that man was made in a single day. The latest findings of anthropology, evolutionary biology and genetics have led to another – no less miraculous – conclusion. Over the course of 450 million years, humans emerged out of the relentless interaction between countless living things and their environments. We share a common origin with all living things on this earth and are therefore genetically related (albeit to varying degrees) to insects, reptiles and mammals, and even to algae, palms and fruit trees. You might say our responsibility for the environment is written into our genes.

For 450 million years, all living things have striven to adapt themselves as well as possible to the conditions in which they live, in order to survive and reproduce. Two conditions must be fulfilled in order for this process to succeed: great diversity within a species, and a set of genes which is subject to constant change.

Changing genes, diversity within our species and the attempt to live in harmony with our environment are both the foundations of evolution and the basis of human existence. Our genetic material is put together in a new way every time a child is conceived, and that means every one of the world's nearly 8 billion people is unique. Every human spends their whole life adapting to the many and varied demands of the environment, in a way that will allow them to satisfy their needs to the greatest possible degree. This effort to live in harmony with the environment is at the very heart of the Fit Principle.

Modern humans are the only living creatures to have developed an irresistible and constant drive to develop their abilities and expand their knowledge. Our aim is not just to understand our environment, but to make use of it and ultimately gain mastery over it. The effort to achieve harmony with the environment has become a need to dominate the environment. Over the past 200 years, scientific, technological and economic progress has accelerated exponentially. There have been far more innovations in the last few decades than in the whole of human history before that period – and they have brought great rewards, but also increasingly worrying consequences for the environment and for ourselves. We no longer live in small communities, as our forefathers did for 200,000 years, but in an anonymous mass society.

Questions we will consider here:

- How can the incredible diversity among humans be explained? And why do all humans have a common genetic code, despite this diversity?
- How much does our genetic makeup change from one generation to the next?

- How did our cognitive, linguistic and social abilities develop? Where does our insatiable thirst for knowledge come from?
- What lies at the root of our irrepressible need to control our environment? How do we prevent ourselves from destroying life on earth, and ourselves along with it?

CHAPTER 2: THE COMBINED EFFECT OF GENETIC PREDISPOSITION AND ENVIRONMENT

*What our genotype accomplishes,
our environment cannot achieve – and vice-versa.*

What applies to evolution on a macro level also affects our own development on a micro level. From birth to old age, our lives consist of the constant interaction of our genotype (the sum total of all our genetic material, combined in a way that is unique to each of us) and our environment. And so we wonder: what part of our essential being is inherent or innate, and what do we acquire? It's a question that intrigues scientists and laypeople alike. Roger Federer is one of the most successful tennis players of all time. Why has he pulled off twenty Grand Slam victories? Because he has been blessed with an extraordinary talent, because he has trained a great deal, or because talent and the desire to train came together in him to form an ideal combination? If parents are particularly empathetic and caring with their children, is their behaviour rooted in a high innate social competency, or was that caring behaviour instilled in them as children? If young people devour a huge Harry Potter book in the space of a week, while some of their schoolmates have trouble deciphering a short column in a tabloid newspaper, is it because their innate reading abilities are miles apart, or because they've received different levels of support at home and school – or are both of these things true?

The importance we attach to genotype and environment respectively is also relevant to society. What is our attitude, for example, to equal opportunities in education? Do children achieve such different degrees of academic success because they have different levels of potential, or because they receive different levels of encouragement at school? How do we create fairness in the economy when people with such different abilities have to meet the same demands? Do we attribute a great performance to talent, a good education or an exemplary attitude to work? What should we reward: talent, hard work or success? We behave differently as parents, teachers, employees and citizens depending on the significance we attach to genotype and environment.

The important questions to be considered here:

- What portion of our characteristics and abilities is inherent? What do we mean by our genetic predisposition, or genotype?
- What portion of our characteristics and abilities is acquired? What do we mean by the environment?
- What is an individual's potential for development, and where do their limits lie?
- How must our society and economy be constructed in order to do the greatest justice to the diversity of needs and talents among humans?

CHAPTER 3: DEVELOPING INTO UNIQUE INDIVIDUALS

Curiosity is the driving force of development.

Every child's development recapitulates a stretch of evolution – on fast-forward, so to speak. Children are born with a huge potential for development, which has evolved and been put to the test over many hundreds of thousands of years. They want to realize this potential. Just a few months after birth, a child starts to grasp objects and understand simple causal connections. At a year old, he can walk unaided and understand a few words. At three, he starts drawing and building houses out of Lego bricks. At five, the child's speech is largely error-free, and he has a basic understanding of numbers. Then he goes to school, and between that point and the end of puberty his abilities take another quantum leap.

When this child begins to grip, to speak, to read and do sums, an exceedingly complex process of maturation is taking place in his brain, which can only be achieved if he is permitted to have the necessary experiences. To this end, he is equipped with a boundless curiosity and a genuine love of learning. He can't help but be interested in every aspect of his environment. He wants to encounter the world, to understand it as well as he can and to prove himself within it.

Insights into child development aren't just helpful for supporting a child in his development. They are also a wonderful point of access for improving our understanding of ourselves, and how we became what we are now. Why some of our abilities are so well honed, and others much less so. Why we have such great interest and such an astounding willingness to learn in some areas of life, and hardly any in others.

Important questions:

- What does brain maturation contribute to development? How significant are our experiences with our social and physical environment?
- What do we mean by curiosity and willingness to learn? How does a child acquire abilities, skills and knowledge?
- What forms can learning take? What does child-centred, lasting learning look like?
- What are we able to learn as adults, and what can we no longer learn? How does our adult learning behaviour differ from that of children?

CHAPTER 4: BASIC NEEDS THAT SHAPE OUR LIVES

Each human has a unique profile of needs.

Humans have always shared elementary needs, such as the need for nourishment, with the more highly developed animals. In the most recent phase of our evolutionary development, however, humans have expanded the ways in which we satisfy our needs to such a degree that these needs have taken on a completely new meaning. Humans don't just find food. For thousands of years we have been cooking and seasoning our food, and holding celebratory communal meals on special occasions with elaborate table settings, wine and candles.

From the Fit Principle point of view, our lives are determined by six basic needs. Alongside the *satisfaction of physical needs*, we have a great *desire for emotional security*, as well as for *social recognition* and a secure position in the family, in our circle of friends, the world of work and in society. If we receive enough emotional security and recognition, we feel at ease and accepted. If we are sidelined, we feel rejected and emotionally insecure. Two other basic needs are the *desire to develop our talents* and *to achieve* the things for which our talents suit us. Children in particular have a strong urge to develop their abilities and acquire skills. The final basic need that drives us is the need for *existential security*. A regular income and security for ourselves and our possessions are very important to us. Unemployment, financial worries, or even the loss of our worldly goods and the threat to life and limb can have an extreme impact on our wellbeing.

Our mental and physical state depends on whether we manage to satisfy our basic needs sufficiently. Pursuing this aim takes up all our strength and time.

The following questions arise here:

- What do we mean by 'basic needs'? Where do they come from, and what do they consist of?
- How do our basic needs develop over our lifetimes, and how significant are they in the different periods of our lives?
- What feelings and ideas are connected with our basic needs? What are we trying to express with them?
- How different can these basic needs be among humans?

CHAPTER 5: COMPETENCIES
WE WANT TO DEVELOP

Humans achieve countless things no other living creature is capable of.

Intelligence is frequently equated with intellectual ability and the intelligence quotient. But our mental abilities go far beyond what can be measured by established tests. We need to factor in motor skills, which are essential for craftsmen such as joiners, or for playing a musical instrument, and social behaviour, which encompasses the different forms of interaction between people and the ability to put yourself in another person's shoes and understand their behaviour.

Terms like 'intelligence' and 'intelligence quotient' also suggest a unified performance by the brain. Nowadays, however, we recognize a multiplicity of intellectual abilities. They are set at different levels not only from one person to another, but within individuals. This means there are people who are very talented linguists, but are much less good with numbers. For other people, it's the other way round. And that means a single figure like the intelligence quotient can't do justice to a person's individual profile of talents. This chapter introduces eight talents, or 'competencies'. Each of these competencies stems from abilities such as visual perception, which we have in common with more highly developed animals. Visual experiences give rise to our first ideas about space, then to linguistic concepts such as spatial prepositions, and finally to activities such as drawing or building houses.

Questions to be addressed here:

- What do we mean by competencies? What do they consist of?
- How do competencies develop into abilities, skills and ideas?
- How different can the level of these competencies be from one person to another?
- How different can the level of these competencies be within an individual?

CHAPTER 6: OUR IDEAS AND BELIEFS

Humans are the only living creatures that have to explain the
world to themselves in order to live their lives.

Ideas enable us to think, understand things and use speech. For example,
I am currently contemplating what ideas mean to me, and setting down
my thoughts in these lines. We start trying to understand the world when
we are very young. We create a world from the ideas we form based on
our experiences with our environment. We explain the world to ourselves
almost compulsively. We just can't help it. We can't imagine a life without
ideas. The acquisition of ideas is what makes us into human beings.

We exchange our thoughts and beliefs with other people, and share
common ideas, including those of a religious nature. Over the course
of our lives, we take on some value judgements from the people
around us. These values can exercise tremendous power over us and
shape our lives within a community to a large extent. For centuries in
many countries, the Catholic Church regulated human relationships
with its dogma and morality. It insisted that people accept its ideas on
the roles of men and women, and marriage and divorce. But even
mighty works are abandoned or just lose their significance when living conditions undergo a fundamental change. Today, after more
than 200 years of Enlightenment, people are guided more by secular
than religious ideas, for example when it comes to the equality of
men and women, or attitudes towards homosexuality.

We are guided by our ideas, and use them to justify our actions in
everyday life just as we do in global politics. It is therefore worth
examining the nature of our ideas and the influence they have:

- What do we mean by ideas? What distinguishes thoughts,
  memories, words and mathematical formulas?
- How are ideas formed during a child's development? How do
  our experiences within the family and in educational institutions
  influence our world of ideas?
- What is the significance for society of ideas such as equal
  opportunities? How do they come about? How do they win
  through?

- What is the significance of consciousness for the accessibility of our ideas? What *is* consciousness? Do ideas exist in the unconscious mind?

CHAPTER 7: FROM NATURE TO THE MAN-MADE ENVIRONMENT

> In order to survive, all living things need not just any environment, but one suited to their needs.

For some decades now, we have been very concerned about our environment – and with good reason. Global CO_2 emissions reached a record level of 36 billion tons in 2013, which in the worst-case scenario will lead to a rise of several degrees in global temperatures this century. In the short period between 2000 and 2012, deforestation destroyed a 1100 x 1100 kilometre area of forest and the habitat of countless animals and plants. In a few decades, human cities and areas of habitation will have grown to cover an area the size of Australia. We are plundering the earth's mineral reserves, polluting its waters with chemicals and littering our environment with waste. It is high time we realized our responsibility towards the natural world. But we should ask ourselves not just what we are doing to nature, but how much we are damaging ourselves in the process. How much of the natural world do humans need in order to remain mentally and physically healthy? After all, our forebears spent the past 200,000 years living out in the natural world, not in sterile rooms. We were originally made for a life in nature.

In the space of just 200 years we have largely cut ourselves off from nature and created an environment shaped by scientific progress, technology and economics. This migration has also fundamentally changed the ancient structures of communal life. With the advent of industrialization, our traditional communities began to dissolve. Large families with numerous children and relatives shrank to become small families with one or two children and just a few relatives. Partners and parents increasingly live apart. The modest communities of our ancestors, allied to the natural world, have become anonymous mass societies based in cities.

Do we, especially as children, still feel secure under modern living conditions? As adults, do we still receive the recognition and affection we need? Can we really get by without a stable network of familiar people? Could it be that a lack of emotional security and social recognition has led to a rise in psychological disorders such as ADHD in children and depression in adults?

We must therefore examine not only our interaction with the natural world, but the influence that the environment we have created has on our lives:

- What is the significance of nature for our wellbeing?
- What effect has the transformation of our original communities into an anonymous mass society had on our wellbeing?
- What effect have reduced family structures had on child development? To what extent are adults dependent on reliable partnerships and a stable social network?
- What happens if we can no longer fulfil our emotional and social needs in modern society? What effects will that have on our mental and physical health?

CHAPTER 8: LIVING THE RIGHT LIFE: THE FIT PRINCIPLE

*Being true to our individual nature is a challenge
that keeps us on our toes all our lives.*

For millennia, humans have been trying to give meaning to life through religious and spiritual ideas, the humanities and, most recently, the ideas of neurobiology. And every religion, ideology and theory develops its own ideal image of the human. These ideas are frequently linked to lofty aspirations such as improving human nature or transforming the world into a paradise.

The Fit Principle is not designed to present another of these ideals. The intention is rather to come as close as possible – without an overarching metaphysical or theoretical structure – to the unique nature of individual human beings, and their effort to live the life that is right for them. The principle is based on the following core assumption, which is taken from human evolutionary biology and which governs the everyday life of the individual:

> Every human strives to bring their individual needs and talents into harmony with their environment. The more successful they are in this, the greater their wellbeing, sense of self-worth and self-efficacy.

Of course, we don't always manage to live a fitting life, even if we strive for it day after day. Sometimes this is down to us: we have unrealistic expectations, don't understand our basic needs or don't make the best use of our competencies. Sometimes it's due to external factors in our lives – and often it's both. Again and again, we struggle back to our feet and set ourselves new challenges, to give our lives direction and purpose again. Over the course of our lives we get better at utilizing our strengths and accepting our weaknesses. We become more familiar with our needs and our potential for development, but also with our limitations, and in this way we gradually become our essential selves.

The Fit Principle is not about achieving as much as possible, reaching the highest social status or heaping up the most wealth. If humans could only be satisfied once they had reached the pinnacle of achievement, the overwhelming majority would sink into unhappiness. And that is by no means the case. Most people are satisfied if they can fulfil their individual needs sufficiently and realize most of their talents.

Questions to be answered on the Fit Principle:

- What does a Fit constellation look like? And what effect does it have on our wellbeing?
- How can we create harmony with our environment? What must we contribute to this, and what does the environment need to contribute?
- How can we understand our basic needs, competencies and ideas well enough to recognize our potential for development, but also to accept our limitations?
- How can we support others and help them live in harmony with their environment?

CHAPTER 9: MISFIT CONSTELLATIONS

The Fit Principle is all about tackling the Misfit situation
by questioning your current way of life.

No human manages to live in permanent harmony with their environment. Smaller Misfit situations, which an individual can overcome without any great difficulty, are part of everyday life. They have no impact on our physical or mental wellbeing. Rather, they are a constant incentive to re-examine our habitual behaviour, ideas and ambitions, and adapt to changing circumstances. But if demands on us (at work, for example) go beyond a certain level, which differs from person to person, they create a Misfit constellation that has repercussions. These constellations can make people feel helpless and powerless, and appear tense and anxious. They tend towards aggressive behaviour or become withdrawn. They may suffer from psychosomatic disorders such as stomach complaints, and increase their consumption of addictive substances like alcohol or prescription medication.

Misfit situations affect some people more than others, depending on which basic needs, competencies and ideas are involved, previous experience of Misfit situations, and the stresses and strains of their life in general. For an older person, losing a job can lead to a genuine life crisis, complete with existential anxiety and a feeling of having been devalued, while the stress remains low for a young adult in the same situation, as they have alternative career paths open to them.

There is a wealth of medical, psychological and esoteric treatments available for people going through any kind of Misfit situation. The Fit Principle doesn't just focus on relieving symptoms such as headaches or sleep disorders, but on tackling the Misfit situation itself. How much have I contributed to the current Misfit situation – by, for example, under- or over-estimating my competencies at work? What has the environment contributed, for example by overwhelming me with complicated tasks? What Misfit constellations have I already experienced, how did they come about and how did I deal with them?

Questions raised in Chapter 9:

- What do we mean by a Misfit? How can a Misfit arise? What are the causes behind it?
- How can we recognize a Misfit situation? How does it affect our wellbeing? How does it affect our physical health?
- How do we tackle a Misfit situation? What basic needs are affected? What expectations do we have of ourselves and the environment?
- How should we evaluate our current life situation? How does the environment contribute to a Misfit situation?
- How can we help others who find themselves in a Misfit situation?

CHAPTER 10: CHANGING TIMES

We have to think the impossible.

In an ideal society, a paradise on earth, all humans would be able to live a fitting life. In Fit Principle terms, this society would be set up in such a way that all people could live according to their individual nature. They would be able to satisfy their physical needs, and they would feel secure and provided for in their community. They would be able to develop their talents and achieve things they found fulfilling. They would feel existentially secure and not threatened in any way. And they would be able to live a life they themselves had determined in every respect.

Have we arrived in this paradise? In some respects, we have. The scientific, technological and economic progress of the past 100 years has vastly improved people's mental and physical wellbeing – even if this doesn't yet apply worldwide. In highly developed countries, the health of the population is better than it has ever been, and life expectancy has doubled. People have access to an advanced education system. In much of Europe, material comfort and peace have reigned for seventy years, an unprecedented length of time. And yet a general feeling of satisfaction has yet to emerge. There is a vague sense of unease in people's minds, the causes of which we are gradually starting to see.

One of these causes is the disregard for humans' emotional and social needs. Humans are deeply social animals whose wellbeing depends on a form of communal life that used to exist in earlier times: stable relationships with familiar people and a culture that creates a sense of identity and common purpose. Now, in the course of modern progress, within a few generations small communities have become a gigantic, anonymous society, for which we are not really suited. We find ourselves in constant competition with one another. We have to prove ourselves over and over as partners and workers, and we are always in danger of dropping out of all our relationships and becoming socially isolated. For most people, emotional security is only a temporary state of affairs. We lead our lives as if we could afford to do without enduring and sustainable interpersonal relationships, as if our mental wellbeing didn't depend on them. But this attitude is proving to be a fallacy. An anonymous, highly complex society and economy cannot create trusting relationships or fulfil our basic social and emotional needs. These things can only be done by a community of familiar people who form a reliable and sustainable network of relationships. It is high time we gave some serious thought to how we want to live together in future, but also to how we should deal with the other causes of general insecurity: the threat of mass unemployment, the meaninglessness of work and the loss of cultural values. To do this, we must imagine the supposedly impossible. Only then will we be ready to reshape our society and economy from the ground up, so that people can satisfy their own basic needs and exercise their individuality.

Questions we will address in Chapter 10:

- To what extent are people today shaped by the legacy of the past – for both good and ill? Are we really completely adaptable and suited to every kind of environment?
- How can we reconcile diversity and individuality with values such as equality and fairness? In view of the great diversity among humans, is a fair society even possible?
- How must a society be constituted so that people are able to exercise their individuality without destroying social cohesion?

- How can quality of life be maintained when automation and digitization are putting more and more people out of work?
- Within state and commercial institutions, who is actually responsible for the mental and physical wellbeing of billions of people?
- And the most important question: how can we strengthen the family so that people can recapture the joy of bringing up children? And how can we create new kinds of community, in which people can satisfy their basic needs better than in an anonymous mass society?

The diversity among living things, the unique nature of all living things and their constant struggle with their environment are some of the basic principles of evolution, and thus also of humanity and human nature. They are part of the human condition, and have found expression in religion, philosophy and art for millennia. In my clinical and scientific work, and of course in my own life, I have always been very moved by the efforts of individuals to bring their individuality into harmony with their environment. These experiences lie at the heart of this book.

I

Human Biological and Socio-cultural Development

Humans are related to every living thing on this planet.

> *There is g-randeur in this view of life, with its several powers,*
> *having been originally breathed by the Creator into a few*
> *forms or into one; and that, whilst this planet has gone cyc-*
> *ling on according to the fixed law of gravity, from so simple*
> *a beginning endless forms most beautiful and most wonder-*
> *ful have been, and are being evolved.*
>
> > Charles Darwin (1809–82)[1]

It has always been one of the greatest puzzles for human beings: why do we, the world and the universe exist? The extraordinary naturalist Charles Darwin found no better answer than that already contained in the first book of Moses: God created the world. In 1863, in a letter to the famous botanist Joseph Hooker, Darwin wrote: 'It is mere rubbish thinking, at present, of origin of life; one might as well think of origin of matter.'[2]

The natural sciences have come up with no more convincing explanation to this day. But we do at least have an idea of what happened *after* the Creation, or – in modern terminology – after the Big Bang, 13.8 billion years ago. Astronomers and physicists are developing ever more sophisticated theories of the universe's expansion since that point, and how it separated into matter and energy. According to their findings, the earth was formed 4.6 billion years ago from gases condensing around the sun. The first evidence of simple life forms stems from around 3.8 billion years ago. Their home was the 'primordial

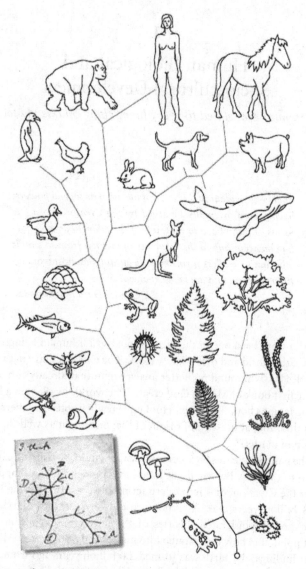

Figure 1.1: (*lower left*) Below the comment 'I think', Darwin made a sketch in his Notebook B of his first idea of the tree of life, 1837; (*right*) a current tree of life.

sea', and traces of them can still be found in fossil form today. In the mid-nineteenth century, Charles Darwin gave a scientific explanation of how life on earth diversified and developed from that point on. His theory of evolution continues to shape our understanding of how life developed. It provided the foundation stone for all discoveries made over the past 150 years in fields as diverse as embryology, evolutionary developmental biology, palaeontology and – above all – molecular genetics. Darwin's theory has been largely proven by researchers in these fields, who have also greatly expanded our knowledge about how all kinds of life forms developed, from bacteria to fungi, plants and animals, and of course humans.[3] The commonly used terms 'family tree' or 'tree of life' are misleading: the lines of descent don't lead off like branches from a trunk, but sprout in all directions like a bush.

The theory of evolution helps us to understand not only where we come from, but also how we became what we are today. It offers an explanation for why humans are related to all living things, such as bacteria and fungi, plants and other animals; how our physical and mental features like the hand or social behaviour developed; and why humans have managed to develop so quickly over what is, in evolutionary terms, an extremely short period of time. The theory of evolution also helps us find answers to the following questions:

How did the enormous diversity among humans come about, and why is each human being unique? Where does the insatiable human drive to understand and control the world come from? And why have humans developed into extremely social animals who also spend their lives attempting to live in harmony with their environment? The initial answers to these questions can be found in biological evolution. It represents the foundation from which socio-cultural evolution, which is what created humans as we know them, was able to develop.

ALL LIVING THINGS DEVELOPED
OUT OF ONE ANOTHER

In 1863, when the British biologist Thomas Huxley first brought Charles Darwin's theory of evolution to public attention in his book

Man's Place in Nature, the idea that mankind might be descended from the apes caused an explosion of outrage. But through his wide-ranging observations of plants and animals, Darwin had made discoveries about the origins of man that went far beyond this. Erring on the side of caution, he kept these thoughts to himself, quite rightly fearing they would cause an even greater uproar and be rejected even more vehemently than Huxley's revelation. Darwin had ultimately come to the conclusion that all living things – plants and animals and humans, too – had a common ancestor.

Fossil evidence

Dating fossilized remains tells us at what point in the earth's history different species of plants and animals appeared, how long they existed for and when they vanished. The oldest bacterial life forms were discovered in a fossilized sea bed. They lived 3.8 billion years ago. The earliest evidence of single-celled organisms with an obvious nucleus (bacteria have no cells) comes from around 600 million years ago. These 'eukaryotes' developed into multicellular organisms at the time when the level of oxygen in the atmosphere began to rise from 3 per cent to today's level of 20 per cent. From this point on, palaeontologists are able to trace the origins of more and more new plant and animal species by means of fossil finds. They have discovered that vertebrates first appeared in the early to middle Cambrian period (540 million to 500 million years ago). Particularly significant finds are those that show transitional life forms, such as the dinosaur Archaeopteryx, a link between reptiles and birds, and the Tiktaalik, a link between bony fish and tetrapods.[4] Well-documented fossils of horse-like mammals illustrate the way in which, over the course of 65 million years, multi-toed browsing animals the size of foxes developed into modern horses, via many intermediate stages.

But research by palaeontologists has not only shown the evolutionary diversification of plants and animals. It has also revealed periods in which mass extinctions took place. In a relatively short period of time, geologically speaking, numerous groups of plants and animals were decimated or even disappeared from the earth altogether. In the

Permian period, around 250 million years ago, an estimated 90 per cent of all animal species were wiped out. At the end of the Cretaceous period, 65 million years ago, there was another mass extinction, probably as the result of a climate catastrophe caused by a huge meteorite strike, a massive volcanic eruption or another as yet unknown event. This catastrophe led to the extinction of the dinosaurs and numerous other animal species. But these mass extinctions, several of which have taken place over the course of evolution, didn't just destroy life; each time, new life was also created. After the last mass extinction 65 million years ago, entirely new species of plants and animals developed, including the ancestors of the mammals and birds we know today. Despite knowing nothing of mass extinction, Lucretius (c. 97–55 BC) captured this essential feature of nature when he wrote: 'Nothing is completely destroyed that we see living today. Nature creates new from old, and the life of the future blossoms in endless flux from the tomb of the past.'

Common features in developmental biology

Several hundred years ago, humans realized that the distinguishing features of living things didn't appear in haphazard combinations. Plants and animals could be grouped together according to their appearance and placed in hierarchies by looking at certain attributes. The Swedish naturalist Carl von Linné (1707–78) created the foundation for a modern botanical and zoological taxonomy with his binary system of classification. Charles Darwin regarded the similarities in how living things are constructed as a strong indication that they all had a common ancestor.

The zoologist Ernst Haeckel, a contemporary of Darwin's, observed that organisms like fish, tortoises and humans are very different once they are mature. But there are some similarities between them in their early embryonic stages. From this, he extrapolated the 'rule of recapitulation': in its development from fertilized egg to adult animal (ontogenesis), each living creature recapitulates the historical stages of evolution (phylogeny). The rule as Haeckel formulated it has been proved wrong, but it does hold true for the early embryonic stages of related species, such as mammals – another piece of evidence that each species of plant and animal builds on earlier species and on organ systems retained from them.

Certain morphological structures in related species of animals point to common ancestors. These 'homologous' features have developed in different directions, depending on the functions they had to fulfil in their environment. The illustration below, from Darwin's era, shows how similar bones in mammals have evolved differently according to the functional challenges facing each species. Useful structures have been strengthened, and superfluous ones have receded or disappeared altogether.

Six million years ago, our ancestors split off from other great apes. Since then, the hands of primates have continued to develop into differently shaped organs with a variety of functions. Orang-utans live in trees, and their hands and feet mainly help them cling to branches and move along them. Their thumbs are therefore hardly developed at all. Gorillas, on the other hand, live almost entirely on the ground, though they can still climb tall trees. Their 'knuckle walk' places their weight on their second and third fingers, which have thus developed into very powerful digits. Chimpanzees, who spend part of their time in trees and part on the ground, have better-developed thumbs – evidence that they can manipulate objects, as when they crack a nut with a stone or piece of wood.

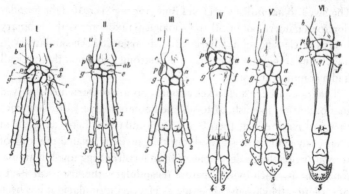

Figure 1.2: Homology of hand bones in various mammals: (*left to right*) human, dog, pig, cow, tapir, horse. (source: Gegenbaur 1870)

Figure 1.3: Development of the hand in primates over the last six million years.

The most important feature of the human hand is its large thumb. It is opposable, which means it can meet each of the other fingers, especially the forefinger. This pincer grip enables us to pick up the smallest of objects. We can lift and carry heavy things using our whole hand, and use large tools such as hammers. We use our hands for a host of physical tasks, but also for interpersonal communication and intellectual activities. We might point a finger in a particular direction to draw attention to something, or wave goodbye to someone. Schoolchildren count on their fingers. Deaf people use gesture, movement and the positioning of their fingers and hands to replace the elements of speech in sign language. A sign also used by hearing people is putting our hands together in prayer. It points to a close relationship between hand and intellect.

Common threads of life

When Charles Darwin published his groundbreaking book *On the Origin of Species* in 1859, followed by *The Descent of Man, and Selection in Relation to Sex* in 1871, the structures and functions of the body's cells and the cell nucleus were still largely unknown, and it would be many years before chromosomes were discovered. At around the same time (1865) the Augustine monk Gregor Johann Mendel reported on his pioneering cross-breeding experiments with pea plants, and the 'Mendelian laws of inheritance' he had derived from them.[5] However, his writings became widely known only forty years later, long after his death, meaning that Darwin had no knowledge of anything approaching modern genetics. But Darwin still noted the diversity and similarities among plants and animals with a keen eye, and was interested in the results of breeding cattle and domestic animals, which he observed on farms and at livestock markets. He too conducted cross-breeding experiments, using pigeons. The successful breeding processes that caused the features of a species to undergo obvious changes in the space of a few generations led him to conclude that a comparable process of selection must be taking place in the natural world. Using nothing more than his extraordinary gift for observation and his highly developed powers of analysis, Darwin grasped the ways in which characteristics were preserved or changed in the natural world, and how living things adapted to their environment.

It was many more decades before the cell nucleus, the chromosomes contained within it and finally the threads of life within the chromosomes, the DNA double helix, were discovered and their significance understood. In 1944, Oswald Avery and his colleagues proved that deoxyribonucleic acid (DNA) was where inherited information was stored. DNA is an extraordinary substance that stores all the information on an organism's structure, development and functions, and passes it on very reliably to the next generation. It has been doing this for hundreds of millions of years, from the first life forms to all the species of plants and animals that have ever existed, including humans.

Molecular genetics confirmed Darwin's assumption: all living things, including humans, have a common origin. As incredible as it may sound, certain sections of our DNA originate in the DNA of the first life forms to populate the earth. We aren't just related to primates, but – albeit to very different extents – to all other living things as well, including fish, the platypus and the chicken. We have genetic material in common with bees, worms and even with grape vines and mildew. As the illustration shows, the protein-coding genes prove a genetic link between humans and the widest variety of animal and plant species. The protein-coding genes are those that trigger the production of specific proteins. These proteins are some of the building blocks used to create and regulate the cells of the body. The closer another species is to humans in the evolutionary family tree, the more genes we have in common with it. The overlap between humans and

8

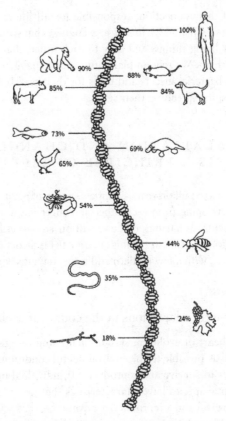

Figure 1.4: A common heritage. The percentage of protein-coding genes that humans share with other living things. (source: Herrero 2013)

chimpanzees, our closest relatives, is estimated at between 90 and 99 per cent of the whole genome, depending on the methods used in the study. Which comes as no real surprise when you consider that our ancestral lines only parted ways 6 million years ago.

Humans, then, are not fundamentally different from the more advanced species of animals – we have merely developed our needs and abilities further, making them more sophisticated than those of any other species (see Chapters 4 and 5). We are genetically related to all living things. This insight makes us, as the only living creatures

9

conscious of this connection, responsible for all life on this planet. All the more so when we're forced to recognize that we are so closely tied to other living things, and they to each other, that our survival depends on them. We rely on plants to produce oxygen, and they in turn rely on bacteria, insects and birds for their metabolism, for fertilization and for spreading their seeds.

ADAPTATION AND CHANGE IS A PRINCIPLE OF LIFE

For more than 450 million years, countless species of plants and animals have been developing from each other and surviving in a huge range of environments. What has made evolution so successful? Charles Darwin also found an answer to this question in his theory of evolution – an answer that, with a few modern additions, remains convincing 150 years later.

Long-term adaptations in the course of evolution

At the very heart of evolution stands each living thing's struggle to adapt as well as possible to the environmental conditions in which it lives, in order to survive and reproduce.[6] If individual members of a plant or animal species have characteristics that give them a greater chance of survival and a reproductive advantage, they will have more offspring – although their offspring won't all inherit these characteristics. The 'natural selection' that arises from this doesn't affect the genes, but rather the characteristics developed by an organism, and it doesn't affect just one individual, but whole groups: plants like wheat, for example, or herd animals like buffalo. Over the course of many generations, advantageous characteristics appear more frequently, and disadvantageous characteristics less so. Organisms adapt to their environmental conditions with increasing specificity, and the frequency distribution of their genes gradually changes. If the genetic differences within or between populations of a species become ever more pronounced and more numerous, a new species will ultimately emerge, the individual members of which can only reproduce among themselves. Humans force this process of selection when they breed plants and animals. The deliberate pairing of dogs has led to the emergence of a tremendous variety of breeds in a very short space of time (from the perspective of evolutionary biology); from a Chihuahua lap dog to a giant Newfoundland or a lightning-fast greyhound. But, as different as dogs may be in their size and shape, they are all still part of the same species.

A key term in evolutionary theory is 'fitness'. And this term, as used in evolutionary biology, has nothing to do with sport or peak physical condition. The quote from Hubert Spencer about the 'survival of the fittest' is also used wrongly in social Darwinism to mean the 'survival of the strongest' – a state of constant competition between the individuals of a species. What the phrase meant to Charles Darwin, however, was 'the survival of those who are best fitted to their environment'. And a huge variety of animals can fit themselves to the same environment.

In order for selection to take place at all, the genetic make-up of a species must be widely varied and subject to constant evolutionary change. The diversity and development of the genome, and the development and differentiation of organs, functions and behaviour that come with these, are further basic elements of evolution. They enable a species to adapt to changing conditions.

In the course of evolution, the genome of all living things has changed slowly but constantly – although only where change has been advantageous. Some areas of the genome are extremely stable, while others change continually. Some genes are retained under varying environmental conditions, while others disappear, change or are re-formed. There are genes and gene complexes that have stayed the same for hundreds of millions of years, such as the Hox gene, which has been determining the structure of body segments and limbs along the body's axis in insects, fish, birds, mammals and humans for more than 400 million years.

There is still very little research available on the molecular-genetic mechanisms that guarantee such a high level of stability in genes. It is worth noting that we have not just one but several copies of some genes. Perhaps nature is ensuring stability in the same way a sensible computer user might: it is making back-up copies. In recent years,

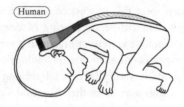

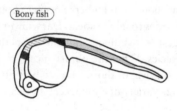

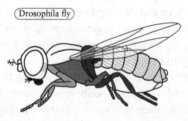

Figure 1.5: The Hox gene has been determining the structure of body segments and limbs along the body's axis in a wide variety of species for more than 400 million years.

scientists have discovered the 'DNA repair gene',[7] the function of which is to ensure that damaged genes are given back their original structure, maintaining their stability.

The long-term evolutionary change of the genome is down to 'mutations'. Mutations can occur in genes and gene complexes as well as in chromosomes. If the former are affected, then a gene's building-blocks, its 'base pairs', are doubled, replaced, changed or destroyed. Mutations can occur spontaneously (at conception, for example) or are triggered by external factors such as radioactivity or chemicals. They can be useful or harmful, or have no effect at all on the gene or the gene complex's ability to function. If mutations occur in the body's organs they can lead to malignant tumours. These 'somatic' mutations are confined to the affected cells and are not inherited. But if mutations take place in stem cells, from which the sperm and the egg develop, they can be passed on to offspring. The development of the genome in the course of evolution is caused by these 'gamete' mutations.

A mutation in the chromosomes can be a 'numerical' chromosome abnormality, producing a change in the number of chromosomes. People with Down's syndrome have three copies of chromosome 21, for example. Individual chromosomes can also undergo structural changes, as when a section of the chromosome is missing or doubled (deletion or duplication). Children with Cri-du-chat syndrome have learning disabilities and a very high-pitched cry, caused by a deletion on the short arm of chromosome 5. The effects of chromosomal disorders are rarely absent and almost never positive.

But the idea that spontaneous mutations and natural selection are the only things that lead to 'meaningful' development of the genome no longer seems plausible. Other mechanisms must exist – particularly those that lead to a kind of linkage between organism and environmental influences. Such a function could be fulfilled by the DNA repair gene mentioned above, for instance. This gene effectively carries out a process of selection among genetic mutations. Mutations that are advantageous for the organism are retained, while those that impair the organism's ability to function are reversed. This means the genome is developed less by spontaneously occurring mutations than by deliberate repair processes. This theory is supported by the molecular genetic research of the past few years, according to which far more mutations occur than previously thought: we simply don't spot them because they are then reversed.[8] 'Epigenetic' processes could be something else that create a connection between genome and environmental factors. For several years now, epigenetics has been looking into the question of how environmental factors can influence the activities of genes on a molecular level. It has proved, for instance, that certain external factors lead to a change in gene activity, by causing a DNA methylation of particular base pairs in a gene, or a modification of the proteins around which the DNA of a chromosome is wrapped (histones).[9] These external factors play an important role in the development of an embryo, and probably continue to have an effect after birth. It is entirely possible, though difficult to prove, that epigenetic changes retained over generations can lead to advantageous, or occasionally disadvantageous, changes in the genome. It is highly likely that there are further exchange relationships between genome and environment about which we still know nothing.

Change from one generation to the next

Changes in our genetic make-up don't just take place along the grand timescale of evolution, but also in the lifetime of every individual human. In choosing a partner, we determine the selection of genes that will be passed on, and during conception a child's genes and chromosomes are put together in a unique combination that has never existed before and never will again.

Sexual selection is an ancient and widespread form of selection across many species of animals. A particular characteristic – the configuration of the tail-feathers in a male peacock, for example – gives the individual with that characteristic an advantage when it comes to reproduction. Sexually attractive males or females reproduce more successfully. In each generation, they pass on their genes more frequently than their less attractive and therefore less successful competitors. The animal kingdom has an incredible repertoire of characteristics that feature in sexual selection. These may be physical features like the size of antlers for stags, the volume and length of song for songbirds, or the body language in the grey goose's courtship dance. Human courtship behaviour covers a very wide variety of factors:[10]

Factors involved in sexual selection (adapted from Morris 1987)

- Physical attractiveness: development of secondary sexual characteristics (beard, breasts), physique (men: broad shoulders, muscles; women: wide hips, soft curves).

- Social behaviour: rituals of encounter (e.g. in nightclubs), display behaviour (men: sitting with legs far apart; women: swaying hips).
- Physical contact: all stages and forms, from holding hands and caressing to intercourse.
- Personality traits and abilities: humorous and entertaining character, social competencies, shared interests.
- Existential and social guarantees: women tend to look for economic security, including income and property, social esteem and status.
- Potential for parenthood: interest in family and children, and in bringing up children together
- Cultural and social value systems: moral ideas, social rules regarding choice of partner and marriage, significance of family and children, group selection on the basis of certain cultural characteristics (e.g. caste system)

For humans, choosing a partner isn't just a matter of erotic charisma and sexual attractiveness. A series of factors such as personality traits, cultural values and expectations of existential and social security also play a fundamental role and contribute indirectly to the make-up of a child's genome. In addition to this, partners often get together because they are similar in certain areas such as intellectual ability ('assortative mating', see chapter 2).

In sexual reproduction, the genome is mixed up twice. A new configuration of genes, gene complexes and chromosomes is first created when the egg and sperm cells are formed in the mother and father. In the process of meiosis, the number of chromosomes in the parent cell from which the egg and sperm develop is reduced from forty-six to twenty-three. The child therefore inherits a particular selection of chromosomes from the mother and father (interchromosomal recombination). The process of intrachromosomal recombination involves the genes on the chromosomes being reconfigured once again: each pair of matching chromosomes, for example the two copies of chromosome 12, is aligned. There is an exchange of genes and gene complexes between the two chromosomes ('crossing-over'). The chromosomes that are passed on to the child are therefore no longer completely identical with those of the mother and father. At fertilization, the two simple sets of chromosomes from the mother's egg cell and the father's sperm cell are united, and a full set of forty-six pairs of chromosomes is formed once again, enabling a child to develop.

A mother and father, with their twenty-three pairs of chromosomes, can each produce 8.39 million (2^{33}) different gametes. When the two sex cells fuse at fertilization, there are 35 billion possible combinations ($2^{33} \times (2^{33} + 1)/2 \approx 3.5 \times 10^{13}$). The probability of a couple having two genetically identical offspring is therefore extremely slight, except in the case of monozygotic (identical) twins. This enormous number of possible combinations also tells us that children can vary in how much they differ from their parents (see Chapter 2). The sum of this difference will probably be an even greater number, since the calculation above doesn't take into account intrachromosomal recombination.

The diversity of the genome is so great that each person is genetically different from all the other 8 billion people on the planet. We are aware of our own uniqueness, and have to deal with our own individuality and the enormous diversity among our fellow humans on a daily basis – which can be a very challenging task.

A masterplan for development

If you are building a house, you don't just need materials; you also need an architectural plan and a timetable detailing the order in which the work is to be carried out. The development of a living thing works in exactly the same way. It requires building materials such as carbohydrates, proteins and trace elements, and a detailed masterplan setting out when each gene and messenger substance (such as enzymes and hormones) needs to be activated in the body's cells, so that the building materials are used correctly.

The code was cracked in 2001: scientists had managed to unlock 99.9 per cent of the human genome, meaning that the DNA sequences of all the chromosomes are now known.[11] Since then, the genomes of plants and animals have been decoded at an ever-increasing rate. As the DNA has been sequenced, researchers have been astonished to discover that the quantity of DNA has not necessarily increased over the course of evolution. The number of base pairs in humans is 3.29×10^9; in a water flea 2×10^8, in the common newt 2.5×10^{10} and in cabbage 7×10^8. There is, then, no direct correlation between the size of the genome and a life form's level of sophistication. Nor does the number of chromosomes reflect how advanced a life form is. The cell nucleus holds 46 in humans, 48 in chimpanzees, 12 in squid, 80 in the blackbird, 48 in the cyclamen and 216 in the common horsetail.

It would also be incorrect to assume that unlocking the human genome means that we now know what humans are made of: that isn't the case at all. Even if we know how many letters are contained in a book, and the order they're written in, we still don't know what the book is about. For that, we have to understand the language in which it is written.

The genome – as mentioned above – consists of several billion base pairs. It is mind-boggling to think that such an immense quantity of information has been passed on so reliably over millions of years, in such a complex process of exchange and fusion. The development of a child, in particular during the first months after conception, is no less a miracle. We know a fair amount about the early development of organs in drosophila, mice and zebrafish, but hardly anything about this process in humans. We may know the sequence of the gene complexes on the forty-six chromosomes, but we have very little understanding of how or when the countless interactions between the genes in the body's cells take place. Millions and millions of cells divide, differentiate and specialize in our organs over the course of days, weeks and months. In the brain, for instance, cells mature into very different types and perform very different functions. It will be decades before we have a real understanding of the highly complex activities of genes in early child development – if we ever do.

Recent studies in molecular genetics confirm that for decades our idea of the connection between genes and the manifestation of characteristics has been too simplistic. Initially, it was thought that the protein-coding genes were much more important than they really are. Humans have far fewer of these than expected: 23,700. Simple animal and plant species have more coding genes, such as the water flea with 31,000, or the cabbage plant with 100,000. Scientists have had to recognize that there is no correlation between the number of protein-coding genes and the complexity of an organism. We share up to 99 per cent of our DNA sequences with our closest relatives, chimpanzees, but our physical and mental characteristics are clearly different from theirs. It cannot be down to the coding genes alone.

The protein-coding genes and gene complexes make up just 2 per cent of the whole genome, while non-coding genes make up the other 98 per cent. For a long time, the latter were referred to as 'junk DNA', because their significance for developmental biology was not recognized. This is surprising insofar as nature doesn't produce anything superfluous. Everything has some function, even if we don't understand what that function is. We are slowly starting to grasp the fact that the non-coding genes are the real key to understanding human development. Some of these genes, with the help of additional nucleic acids such as the ribonucleic acids, programme the development of the unborn child.[12] They determine which genes are switched on and off, and when. The miracle of human development isn't performed by a rowdy mob of genes; this is a highly complex, genetically predetermined process that regulates the spatial and chronological interplay of coding and non-coding genes and their products, such as enzymes. In the course of development, gene activities become increasingly fixed and limited, and the cells become more strongly set in their ways. After the sperm and egg have fused together, 'totipotent' stem cells are formed, from which all the different kinds of cells in the body develop. Over the course of a pregnancy, the cells start to differentiate themselves into types specific to different organs such as the heart, liver and brain.

The key to human development therefore lies less in the size of the genome than in its masterplan. It consists on the one hand of a kind of record of how human evolution has proceeded up to this point, and on the other of instructions for how to build a human being. The masterplan is like the choreography for a ballet, and the genes are its dancers. The number of dancers does as little to determine the type of performance as the number of genes does for the developing organism. And the size of the ballet company also gives no clue as to what piece the dancers will perform. The same ensemble can stage quite different pieces. It is only when the dancers start to move, to interact with each other and create shapes and scenes that the subject and style of the performance become apparent. The choreography prescribes each move made by the dancers during the performance in minute detail. They are active at different times as soloists, in groups or as an ensemble, and become ever more strongly fixed in their function. And the same goes for genes. They are switched on according to a highly complex plan which covers both time and space, in order to set specific processes in motion; then they in turn switch on other genes, and are switched off again. Dancers can stumble, miss each other, forget a sequence or pause for a moment if they are momentarily put off, say, by a sneezing fit in the audience. Genes, too, can become active at the wrong moment – or, for a variety of outside reasons such as viral infections, they may not follow the masterplan to the letter. Fortunately these missteps are very rare.

The development of a human being doesn't take an hour or two, like a ballet; it takes nine long months of pregnancy to produce a child capable of life, and another fifteen years to produce all the characteristics and abilities that go to make up an adult human. The masterplan creates the thing that differentiates a human from all other living things: their genotype. When I speak of the genotype in the chapters that follow, I don't mean the human genome, but the sum total of what makes up the unique organism formed during pregnancy. This organism develops in harmony with the environment each foetus finds in its mother's womb, and continues to develop for many more years through constant interaction with its environment.

Related but different

A few years ago molecular-genetic studies showed Irish and Scottish people that, much to their astonishment, they weren't just descended from brave Vikings from the freezing North, but also from fearless Tuaregs from the hot Sahara.[13] Each of us has distant relations in Asia, Australia or America. Europeans have genetic similarities with Mongols; Indians have similarities with the Maya peoples; and Aborigines with Scandinavians. Genetic studies of populations have proven without doubt that all humans are related to each other.[14] There is no evidence to suggest that human ethnicities or races are based on particular genetic characteristics. There is no single gene or gene complex that clearly divides certain groups of people from others, or gives them advantages or disadvantages with regard to specific characteristics and abilities. Many people find it understandably difficult to believe that there is no specific marker for a particular race, even though we can readily tell people of European descent from Africans, Arabs and Chinese people. Even though people have a wide spectrum of skin tones and blue, green or brown eyes; even though some people have almond-shaped eyes, and some don't.

But how is it that all people are related but different? Most characteristics are determined not by a single gene, but by several; at least three genes are involved in eye colour, for example. In addition to that, a gene can appear in different variations, called alleles. The 'gey' gene consists of two alleles, one for green and one for blue eyes, and 'bey2' consists of one gene for brown and one for blue eyes. And an allele can be either dominant or recessive, meaning that it has a stronger influence over a characteristic than other alleles, or is suppressed by them. The allele for brown eyes is dominant over the alleles for other eye colours. There is a complex mechanism behind the great variety of eye colours in humans. With most characteristics, like the shape of a person's eyes, even more genes with various alleles are involved. For a characteristic such as height, the number of genes and their alleles is so large that we see a continuum in how this characteristic appears (multifactorial inheritance). People aren't either short or tall; there is a steady gradient of height from shortest to tallest.

But this still doesn't explain why characteristics like skin colour, eye shape and height are distributed so differently from one part of the world to another. This is the result of natural selection, which is how we adapt to different living conditions – the climate, for example. The skin colour that is most advantageous to a particular geographical region gradually comes to dominate there. In regions that get a lot of sun, such as Africa, people are dark-skinned; pigments (melanin) protect their skin from excessive ultraviolet rays. Light skin is more advantageous in places like northern Europe, where the sunlight is weaker, because it aids vitamin D synthesis and bone growth. Scientists have been able to prove a correlation between skin colour and geographical region: the higher the UV radiation, the more pronounced the pigmentation.[15] Typically, almond-shaped eyes are seen not only in the Asiatic population, but also in the Sami people of Finland and the indigenous populations of Africa and South America. Scientists suspect that in the early period of *Homo sapiens*, people with this eye shape adapted better to regional weather conditions that included strong sunlight, snow and rainstorms – which often come with poor visibility. Living conditions dictate the selection pressures to which genes and gene complexes are subjected (such as the genes for skin pigmentation in Africa and Siberia), which causes characteristics to be distributed differently between populations.

From experience we know that Japanese people tend to be shorter than Europeans. But this difference in height is due not to different genetic features, but to dissimilar living conditions. Migration and the influence of socio-economic factors prove this theory quite conclusively. Until the 1960s, Japanese children and adults were significantly shorter than Americans and Europeans. But when Japanese people emigrated to the United States and had children who grew up there, these children weren't significantly shorter than American children of European descent. And in the past forty years children in Japan have also grown taller and taller. Both the migration effect and the increase in height in the country of origin can be put down to improved living conditions, particularly a diet richer in calories and protein.[16]

The World Health Organisation (WHO) works on the assumption that the people of all large populations have the same genes for height. The average height varies, if at all, only by a few centimetres between Europeans, Africans, Asians and Australians who grow up in good living conditions.[17] One exception is the Pygmy people. Their short stature is due to a genetic defect in a growth hormone, the 'insulin-like growth factor'.[18]

Essentially we can assume that, for any given characteristic, under the same living conditions the variety between individuals within a population will be far greater than the overall difference between populations. In Europeans, the difference between the shortest and tallest people is around forty centimetres, and in extreme cases up to sixty centimetres. The distribution is exactly the same within the Asiatic, African or Australian populations. The difference in height between one person and another can be more than ten times the difference between populations.

The same applies to intellectual ability.[19] The average intelligence quotient (IQ) varies by just a few points between populations, when differences in living conditions are taken into account – in particular the quality of education available. But the difference in IQs seen between individuals within a population is more than sixty points.

If there are no 'ethnicities' and 'races' with clear genetic differences, then how did millions of Jews, Roma and other population groups come to be persecuted and murdered because of their 'race' during the Nazi period? Such crimes against humanity are not rooted in molecular-genetic differences, but in socio-economic and cultural realities such as poor governance, language and religion. Features of our social behaviour such as xenophobia also play a role here (though never a crucial one). We will come back to this in Chapters 5, 7 and 10.

WHY HUMAN EVOLUTION HAS TAKEN SUCH AN UNUSUAL COURSE

There are some plant species that have changed very little over long periods of time. These 'living fossils' include horsetails and ferns: evidence of their ancestors goes back 350 million to 400 million years. Living fossils also exist in the animal kingdom, examples being ants, the nautilus and the crocodile. The oldest fossil evidence of ants, preserved in amber, is estimated to be around 100 million years old. Some species of ant have changed very little in the intervening period, in terms of body shape and colony-building behaviour. Evidence of the nautilus's ancestors (cephalopods) goes back as far as 400 million to 500 million years. The nautilus as we know it today is 30 million to 60 million years old. Crocodiles are the descendants of the dinosaurs that populated the earth around 230 million years ago; the species of crocodile alive today have existed for around 50 million years. Even if the living fossils in the plant and animal kingdoms are still developing, they have changed so little that their morphological characteristics allow us to trace them back to their ancestors very easily. But why haven't they changed over such long periods of time? From the point of view of evolutionary biology, the most obvious explanation is that they found their ecological niche early on, and it has been largely preserved to this day. There has simply been no need to adapt to changing environmental conditions. The following question is much more difficult to answer: why have millions of plant and animal species, and in particular humans, developed in such an extreme way?

Human evolution

For a long time, the course of human evolution was far from spectacular. The first mammals from which we originally descended were similar to today's rats, and lived 200 million to 230 million years ago. The oldest fossil finds that can be classed as primates are around 50 million years old. Scientists estimate that the great apes (*hominidae*) originated between 18 million and 15 million years ago. Between 6 million and 7 million years ago, a population of African apes split into two species. Via an unknown number of intermediate stages and branches, one of these species evolved into the primates we know today, such as chimpanzees and bonobos. The other, via a similarly unknown number of stages and branches, developed into humans.[20]

Around 4 million years ago, a species of australopithecines, an early predecessor to humans, began to stand up and walk on two legs. Its arms and hands were then free to handle objects and could be used among other things to employ sticks and stones as tools (pebble tools). The australopithecine brain had grown to around 25 per cent larger than those of other primates of the same period. We can only speculate as to what other achievements these early ancestors of ours were capable of. They were probably comparable with those we can observe in today's chimpanzees, who are also able to use tools.

Between 2 million and 3 million years ago, the growth of these apes' brains began to accelerate. These 'human' ancestors (as the genus name 'Homo' tells us) were now capable not only of using objects as tools, but of adapting these tools for a specific purpose. They chose particularly hard types of stone, struck them against others to work them into sharp-edged tools, and used them as axes, spear tips and stone knives, with which they could, for example, hunt animals and remove the flesh from the bones.

It is likely that 1 million to 2 million years ago several groups of australopithecines and other hominids lived alongside each other all over Africa, and many larger groups of these hominids migrated and spread right across Eurasia. We don't know how they lived, or what their cognitive abilities were. There are clues that they had already developed a nuanced acoustic form of communication. The skulls of *Homo erectus*, which populated the earth around 1.5 million years ago, have an indentation on the inside that may be the imprint of the Broca speech area.[21] A lowering of the larynx characteristic of humans can be seen in *Homo heidelbergensis*, which lived between 500,000 and 200,000 years ago. The lower-lying larynx was coupled with an enlarged nose and throat region, enabling these hominids to produce a variety of sounds. Both these findings make it likely that these early hominids had a highly developed form of acoustic communication, though they don't prove the existence of language – meaning communication with a symbolic character (see Chapter 5). This means we don't know whether these hominids used an early form of language, exchanged experiences or even told their children stories and myths.

The species *Homo sapiens sapiens* (generally known as *Homo sapiens*), from which we are all descended, evolved from one of the various Homo groups around 200,000 years ago – we don't know which. This new hominid was more lightly built than any of the previous species in the Homo genus and, most importantly, had a much larger brain, which gave it entirely new cognitive abilities.

Another Homo group, the Neanderthals, which evolved from *Homo heidelbergensis*, populated the whole of Eurasia 500,000 years ago and were probably much more similar to *Homo sapiens* than scientists once assumed. The oldest Neanderthal graves that have been discovered are in the caves of Qafzeh and Skhul in Israel, and are between 120,000 and 90,000 years old. Burials involving burial objects were carried out between 70,000 and 50,000 years ago (Shanidar, Iraq), and possibly before. This means the Neanderthals must have thought about death and the spiritual realm, which presupposes a certain concept of time. For 30,000 years they shared the earth with modern humans, before dying out around 30,000 years ago for reasons unknown. Their brains seem to have been slightly larger than those of *Homo sapiens*, though this clearly didn't give them any evolutionary advantage. Another recently discovered group of *Homo sapiens*, which had contact with both *Homo sapiens sapiens* and the Neanderthals, is known as the Denisovans. They lived around 40,000 years ago in the Altai mountains in southern Siberia.[22]

Molecular–genetic clues suggest there was a degree of intermingling between *Homo sapiens*, Neanderthals and Denisovans. Our genome contains small parts of the Neanderthal genome, including the FOXP2 gene, which is linked to the ability to speak.[23] A variant of the EPAS1 gene, which is present in Tibetans and makes it easier for them to breathe at high altitudes, is said to come from Denisovans, and points to certain genetic crossovers with *Homo sapiens*.[24] Assuming that *Homo sapiens sapiens* did mate with Denisovans, Neanderthals and quite possibly other hominids, then we may assume that they were more similar in their communication and social behaviour, and in their way of life, than was previously thought.

Our early ancestors and phylogenetically closest relatives existed for hundreds of thousands of years – much longer than *Homo sapiens* have been around. But modern man was the sole survivor. There must be significant reasons for this survival that go beyond finding good living conditions.

An extraordinary development

Evolution takes time. It took the horse 55 million years to raise the height of its withers from 40 to 160 cm, and grow from 20 to 800 kg. But only 3 million years passed in the evolution of *Homo sapiens* before its brain had more than tripled in weight, going from 400 to 1,400 g in men and 1,300 g in women. This development is highly unusual in evolutionary biology – and it is what enabled the socio-cultural evolution of humans to take place.

Between 3 million and 6 million years ago, the brains of our earliest ancestors were about the same size as those of the other great apes. Between 2.5 million and 3 million years before our time, their brains began to grow to a remarkable size and – as anthropologists have found – went on growing until about 200,000 years ago. The brain grew from 400 to an average of 1350 g. The brains of our nearest relatives the primates, by contrast, experienced an insignificant level of growth during this time. Why did the brains of at least one

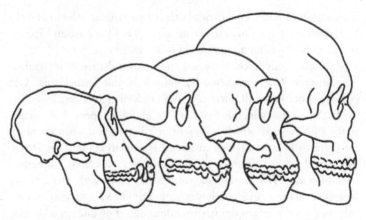

Figure 1.6: The skulls of a macaque, orangutan, chimpanzee and human.

species of *Australopithecus* suddenly start to grow faster than the others, even though there was no corresponding increase in its height and weight?

The first clues to what might have caused this brain growth have been provided by recent discoveries in molecular genetics.[25] The HAR1 gene is ancient and extremely stable. Although the genealogical lines of chimpanzees and chickens split off 400 million years ago, the only difference between their versions of this gene lies in two of its 118 base pairs. The genealogy of chimpanzees and humans split only 6 million years ago. But in that comparatively short period there have been eighteen changes in the 118 base pairs of human HAR1. If the HAR1 gene is absent, a child will develop a 'smooth brain', lacking the furrows of a normal brain (lissencephaly). This defective brain structure causes learning disabilities. Another gene, ASPM, has an even more direct connection to brain size. If this gene is absent, the brain doesn't develop to the right size (microcephaly). The cerebral cortex is largely missing, and the brain weighs only 400 g. It is similar to that of a chimpanzee or gorilla. Without the ASPM gene, we find ourselves back at the stage of evolution when accelerated brain development began, around 3 million years ago. The noteworthy thing here is that both genes, HAR1 and ASPM, are regulatory genes, not protein-coding genes. As an embryo develops, they play a crucial role in shaping and structuring the cerebral cortex. Alongside these two, there are sure to be other, as yet unknown, genes and gene complexes affecting brain development.

As the brain's weight increased, the number of nerve cells rose to 20 billion, and the number of synapses (points of contact between the nerve cells) rose to an incredible 400 billion. Even so, the growth of the brain alone doesn't explain the continued development of intellectual abilities in humans. The brain of an elephant weighs in at an impressive 4.6 kg – about three times as much as the human brain. Elephants are social animals and communicate with each other, but they have nowhere near the same cognitive and linguistic abilities as humans. Alongside quantitative factors such as the increase in numbers of nerve cells and synapses, qualitative factors must have made an important contribution to man's cognitive capability. Abilities like symbolic thinking and language, which set man apart from other animals, require sophisticated, conscious ideas that can only be produced by a

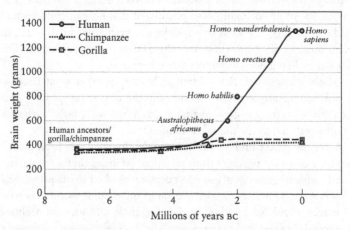

Figure 1.7: Development of brain weight in the early ancestors of humans, gorillas and chimpanzees over the last 7 million years. Since this is based on the measurement of individual skulls, the spread of brain weights is not known. The results must therefore be interpreted with caution (average and range of brain weights for modern humans: for male brains 1,400 ml (range: 200 to 1,600 ml); for female brains 1,300 ml (range: 1,100 to 1,500 ml)). (source: adapted from Pike et al. 2012)

high level of networking between the areas of the brain (see Chapters 3 and 6). How such extreme networking between neuronal structures came about in the first place, and what kind of brain architecture and neuronal interconnection lies at the root of our brain's enormous capability, remains a mystery. Why should the brain have stopped growing 200,000 years ago? Some anthropologists even claim that brain weight has begun to decrease again in the past 28,000 years. But, strictly speaking, we can't rule out the possibility that our brains are still growing. The average increase in brain weight over the last 3 million years has been around 30 g per 100,000 years. The small amount of data we have for the last 200,000 years neither proves nor rules out an increase or decrease of 60 g. There aren't enough measurements available to give us a reliable average.

The developmental neurologist Heinz Prechtl reasoned that the brain had ceased to develop because the head of an unborn child couldn't grow any further: a larger head wouldn't be able to pass through the narrow gap in a woman's pelvis during childbirth.[26] The zoologist Adolf Portmann even postulated that human children had to be born early – a kind of physiological premature birth – due to their large heads.[27] But there are other, far more compelling reasons for babies being born immature (see Chapter 3).

Might it be possible that the human brain stopped growing because it was sufficiently equipped to manage the challenges of life? And could it be starting to grow again now, as humans are faced with new challenges such as the digital revolution?

Over the past 3 million years, brain development has been placed under a constant pressure from selection. Abilities that were particularly advantageous were more frequently passed on to offspring than others were. Good spatial orientation in the natural world was particularly useful for hunting. Social, linguistic and musical abilities improved cohesion in the community. Highly developed logical thought helped people get a better grasp of causal connections – the effect of soil conditions and weather on the growth of plants and animals, for example. An understanding of temporal connections was helpful in gathering fruit that ripened at different times. Gross motor skills used in hunting and fine motor skills for using tools were particularly beneficial. As with all other living things, individual abilities developed differently in different humans. Some people possess more advanced social and linguistic skills, while others have excellent figural and spatial abilities, and others still have particularly well-developed motor skills.

Masters of adaptation

In order to learn about the origins of *Homo sapiens* and the species' migratory behaviour, molecular genetic studies have been carried out in population groups on every continent. The scientists behind these studies focused on certain properties of the mitochondrial DNA and the Y chromosome, and made some extraordinary discoveries.[28]

Modern man is descended from a group of just 10,000–60,000 humans, who lived between 150,000 and 200,000 years ago in East Africa.[29] The 8 billion people in the world today come from a relatively small original population. Around 7500 generations have come and gone since our direct ancestors left the savannah and spread out across the globe. All the Africans, Americans, Australians, Chinese and Europeans living today are separated from early modern man by the same number of generations.

Using molecular–genetic methods, we can also reconstruct the migratory movements of our ancestors and when they took place, even after thousands of years. The decisive wave of migration happened around 65,000 years ago. Humans initially settled in the Middle East, and from there they carried on migrating to Europe, or spread across Asia all the way to Australia and the Pacific Islands. Around 13,000 years ago, several waves of migrating humans crossed the Bering Strait to North and Central America, and eventually pushed on to the southern tip of South America. They didn't move freely all over the land mass, tending instead to follow the rivers and coastlines. And humans certainly didn't take the world by storm. A rate of one kilometre per year was enough to take them all the way round the world over 40,000 years. Our direct ancestors managed to populate every continent – with the exception of Antarctica – and used their extraordinary cognitive and social abilities to adapt to an incredible array of environmental conditions, from the jungles of Africa to the steppes of Siberia and the islands of the Pacific. Around

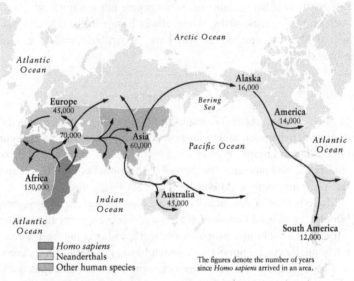

Figure 1.8: The spread of modern man over the past 65,000 years, using data taken from studies of DNA markers.

12,000 years ago, humans began to intervene in their environment more and more. This was the start of socio-cultural evolution.

SOCIO-CULTURAL EVOLUTION

By and large, we are still the same animals we were 200,000 years ago. There have been no significant changes to the genome or the brain of *Homo sapiens* over this period (a very short space of time in evolutionary terms). So why didn't our ancestors invent the smartphone 200,000 years ago? Or is it just that we are ignorant of these things because the evidence of what early *Homo sapiens* achieved hasn't survived?

The prelude to cultural evolution

We can see what humans were capable of in terms of culture as early as 40,000 years ago from the realistic pictures scratched, drawn or

Figure 1.9: (*left*) Cave painting (at least 15,000 to 20,000 years old); (*centre*) vulture bone flute (35,000 years old); (*right*) the Venus of Willendorf (28,000 years old).

painted on cave walls and boulders. They depict hunting scenes and a great number of different animals of the period. Famous sites include Lascaux in the Dordogne (France) and Altamira in Cantabria (Spain); other discoveries have been made in Russia, China and Australia.

The humans of that time also carved perfectly formed animal and human figures out of ivory and stone, using sharp-edged flint tools to produce artefacts like the 28,000-year-old Venus of Willendorf. In 2008, a flute made of vulture bone estimated to be 35,000 years old was discovered in the Hohler Fels cave near Ulm in Germany. You can hear what a flute like this might have sounded like by searching for 'bone flute music' on the internet.

Between 10,000 and 12,000 years ago, cultural evolution began to gather pace, with the start of arable farming and livestock breeding in the Middle East. Previously nomadic hunters and gatherers settled down to become farmers. This 'Neolithic Revolution' took place slightly later in Asia and Central America. People sowed and tended various types of grain and cultivated plants. They domesticated animals, the first of these being dogs, sheep and goats, then cattle and – as researchers have discovered only recently – bees. Cave paintings have been found in Laos of people riding elephants and using ploughshares in fields 6,000 years ago. Scientists estimate that farming and cattle-rearing would have enabled up to fifty more people to live and eat on the same area of land than would have been possible in the previous nomadic era.

Settlement led to new ways of life. People no longer had to spend all their time, day after day, satisfying their needs by hunting animals and gathering fruits and plants. They cultivated ever larger areas of land using increasingly intensive farming methods, and laid down winter stores. Bringing in a harvest of corn or owning a field signified existential security. Even at this time, wealth was probably an expression of foresight and high social status. People moved from simple huts and caves into clay and stone houses, communities grew larger, and the first towns came into being. The town of Göbekli Tepe in southern Anatolia is estimated to be between 8,000 and 10,000 years old.

Sedentary living, farming and raising cattle created unprecedented periods of free time. People no longer had to spend all their waking hours just keeping themselves alive. They were able to make better and better use of their skills, and constantly expand the range of their activities. They invented the wheel, which was first used for turning clay pots and then as a cart wheel. Potters didn't just produce vessels to hold food and drink; their creations were also artistically decorated. And textiles didn't just protect people from the cold and rain; they were also worn as adornment. People began to mine and work with metals. Weapons that had previously been made of stone were now cast from bronze, copper and eventually iron. Jewellery was first made using nuts, shells and gemstones, but in time these were increasingly replaced with precious metals such as bronze, gold and silver.[30]

With over-production of things like cereals came trade, and soon the exchange of luxury goods such as amber necklaces. It is no coincidence that the term 'culture' is derived from the Latin *cultura* (agriculture, cultivation, care). Culture in its widest sense encompasses everything that humans have produced: the visual arts, morality, law, religion, technology, science and business. This 'cultural revolution' isn't coded into our genes like the biological revolution, but is captured in artefacts and documents of all kinds.

Exponential acceleration

One of the best ways to track cultural evolution is by looking at communication. We don't know how long ago our ancestors developed a language with a symbolic function. But it is entirely possible that the

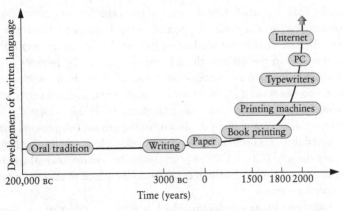

Figure 1.10: Exponential curve of cultural progress, using the example of language. The time axis is not drawn to scale, in order to show progress over the last 3,000 years.

beginnings of language development go back as far as 100,000 years – or perhaps even 200,000. Oral traditions and the social preservation of information may therefore have developed very early in human history. Cultural evolution took a great leap forward with the invention of writing. The first written symbols were used around 5,000 years ago in Mesopotamia. The invention of paper around 2,000 years ago, printing using wooden blocks in the seventh century (both in China), and the printing press in fifteenth-century Europe helped to spread the written word more widely. A veritable flood of information began in the nineteenth century with the mechanization of the printing process. The information technology of the past thirty years has multiplied the volume of data in existence once again, raising it to mind-boggling levels.

Cultural accomplishments, as the graph above shows, haven't simply accumulated over the course of human history; they have grown exponentially. In each time period, progress has been many times greater than in the previous one. Systems of symbols, and the artistic representation of life in drawings and sculptures, developed much more strongly between 10,000 and 40,000 years ago than it did in the preceding 150,000 years. The number of cultural artefacts produced in the 10,000 years before the start of our calendar was many times greater than in the preceding 30,000 years. In the first 1,800 years AD, the rate of cultural production increased many times over once again. And in the nineteenth and twentieth centuries, humans produced a million times more cultural goods of all kinds in the space of just 200 years than in the previous 1,800 years, and even over the whole 200,000 years since the emergence of *Homo sapiens*. The human brain was without doubt one of the crucial components that allowed this exponential development to take place, but it is far from being the sole factor. A series of advantageous environmental conditions and reciprocal developmental processes helped to accelerate our cultural progress:

Improved living conditions. Until around 12,000 years ago, our early ancestors spent most of their time finding food and protecting themselves from extreme weather and all kinds of other dangers. The last ice age ended only 12,000 years ago. Large parts of Europe, Asia and North America lay under a sheet of ice several kilometres thick. Then the climate began to improve, making a vital contribution to the rise of arable farming and cattle-rearing. Humans also found land in Eurasia that was much better suited to farming than the jungles, savannahs and deserts of Africa.

Settlement. Once people became settlers, the millennia that followed saw their activities becoming more diverse and more specialized, for instance through the use of ploughs and three-field crop rotation. There was an increasing division of labour, in the work of craftsmen, for example. In smaller tribes, skills that had been learned by one generation frequently died out again, but in larger communities these skills were handed down, developed further and exchanged with other communities.

Food. Growing food and raising livestock meant more people could be fed. There were still periods of hunger, but thanks to the practice of laying down stores these were less frequent and less devastating than in earlier times. The Pharaoh's dream of the seven fat and seven lean years in the First Book of Moses illustrates the existential significance that precautionary cultivation had gained even at that time. Anthropologists have suggested that, although people had more to eat as settlers, their diet wasn't as good as their nomadic ancestors' had been. Food improved in relation to its calorific content and the provision of carbohydrates and protein. But in comparison with the eating habits of the hunter-gatherers, the quality of nutrition fell. It was less varied, and often so unbalanced that people began to suffer from anaemia, deformations of the spine, and vitamin and mineral deficiencies.[31]

The invention of writing. Settlement led to an increase in trade between communities. The more goods were exchanged within and between the settlements, the greater the need to measure and record quantities, numbers and areas. Writing and counting systems were originally invented for this purpose. An early form of writing – symbolic marks on earthenware jars – was developed around 5,500 BC. It came from the Vinča culture, which was native to modern-day Serbia and Transylvania. The first recorded real writing system, known as cuneiform, was invented by the Sumerians in Mesopotamia around 3000 BC. It is made up of symbols scratched into clay tablets with a stylus. Pictorial hieroglyphs were developed in Egypt around 3000 BC. These early forms of writing were developed further by the peoples of the Middle East, the Phoenicians, Egyptians, Greeks and Romans, finally becoming the Latin alphabet we use today.

Writing systems didn't just make trade quicker and easier. They allowed people to record the laws and decrees needed to regulate life in increasingly complex communities. They also made it easier to preserve and pass on customs, stories and religious ideas, and so writing increasingly replaced oral traditions. The *Epic of Gilgamesh*, one of the oldest surviving written poems, comes from the heyday of Babylon, and is around 3,500 years old. Even then, it told of a great flood and a proto-Noah.

The invention of numbers. We don't know when humans began to count, but it was probably tens of thousands of years ago. Numbers and sums were first recorded in written form by the Sumerians, in around 3000 BC. The oldest examples of algebra are found on an Egyptian papyrus document from 1700 BC. It contains the following problem: 'An amount added to one seventh of itself equals nineteen. How large is the original amount?'

The replacement of Roman numerals with Arabic numbers led to a breakthrough. The latter use a decimal system that makes carrying out mathematical operations far more efficient than the categorical Roman system. They also contain the mathematical concept of zero.

Education. Education in the sense of a deliberate schooling in academic disciplines such as reading, writing and mathematics, and the imparting of cultural accomplishments, began around 4,000 years ago, but for a long time it was reserved for a small, elite circle of people. General compulsory education was introduced only in the nineteenth century, when society recognized that scientific, technological and economic progress necessitated raising the general level of education in the whole population.

Health. The first piece of writing dealing with health and sickness once again comes from ancient Babylon. The Code of Hammurabi was written in around 1800 BC and provides information on the use of medicines and magic substances. We can assume that people have been thinking about health, sickness, dying and death for tens of thousands of years. They studied the use of medicinal herbs, and worked as midwives and religious healers. From the ancient world to the Renaissance, the healing arts were ruled by the doctrine of the 'humours'. Modern medicine emerged around 400 years ago, from a union of the natural sciences and the healing arts. Andreas Vesalius studied human anatomy, in particular the heart, and William Harvey was interested in how the circulatory system functioned. But real progress in medicine was only seen towards the end of the nineteenth century. New technological equipment such as the high-resolution microscope enabled biologists like Louis Pasteur to prove that bacteria were the cause of diseases including cholera, plague and tuberculosis. Biochemists studied the metabolism and discovered messenger substances like insulin and enzymes such as lactase. Wilhelm Conrad Röntgen detected and made use of X-rays. Working with discoveries made in the natural sciences, a fast-growing pharmaceutical industry developed vaccines, narcotics, antibiotics and countless other drugs.

But the significant improvement in human health was not only due to accomplishments in medicine. Hygiene measures such as clean drinking water, toilets and sewer systems, an adequate supply of high-quality foodstuffs and much improved housing also made an important contribution.

Raw materials. Our ancestors discovered minerals (salt, precious stones) and metals (copper, zinc, iron) in the earth. Raw materials were so important for cultural development that archaeologists named periods such as the Bronze Age and the Iron Age after them. Over thousands of years, humans developed increasingly efficient methods for mining natural resources, extracting them from earth and rock, processing them and using them in a huge variety of ways.

The use of energy sources. The earliest fire pit, at more than a million years old, was discovered in the Wonderwerk Cave in South Africa, and another fire pit with burnt food residues was found in Gesher Benot Ya'aqov in Israel. The latter has been connected to *Homo erectus*. Fire must have been essential to the survival of *Homo heidelbergensis* as they colonized Europe north of the Alps 600,000 years ago – at the time, an extremely cold region. Roasting meat and cooking plants opened up new sources of food to our ancestors and improved their nutritional intake. At the same time, fire offered warmth, light and protection from predators and insects. Eventually, it also allowed them to exploit materials such as metals and clay.

With the advent of industrialization, the demand for energy grew enormously. It was initially met by steam power, and later through the exploitation of oil and coal deposits. But it was the discovery of electricity that brought the great breakthrough in the supply of energy. Energy could now be transported over great distances at little cost, and was available at all times. In the twentieth century, the constantly growing demand for electrical energy led to the use of hydroelectric and nuclear power and, more recently, solar and wind energy.

Industrial Revolutions. Each stage of industrial progress has been triggered by scientific discoveries that were then turned into technology and used for commercial purposes. The discovery of the laws of mechanics led to the mechanization of work processes in the eighteenth and nineteenth centuries. From then on, physical labour was increasingly replaced by machines. The discovery of electricity in the nineteenth century allowed the development of entirely new communication systems such as the telegraph and telephones, and in the twentieth century the invention of radio and television. And in recent decades digitization has succeeded in transferring analogue data into discrete values that can be saved, processed and disseminated electronically. An estimated 98 per cent of information used globally today is in digital form. The digital revolution has accelerated the production of goods by robots of all kinds and is now leading to new services such as online shopping, which will change our lives profoundly in the decades to come. It has also given billions of people access to the virtual world of the internet.

Cultural evolution can best be described through the many inter-actions between humans and their largely man-made environment. The following forms of interaction are key to the process of cultural evolution:

- With constantly increasing knowledge and increasingly capable technology, humans have learned to make better and better use of the environment's resources (agriculture, raw materials, energy).
- Discoveries and achievements help to spark creative ideas, which in turn lead to technological advances and commercial products (the 'ratchet effect').[32] This spiral of discovery, technological application and commercial exploitation turns faster and faster over time.
- Humans have created an environment in which they can both preserve their knowledge and achievements and constantly extend them, applying them ever more efficiently to activities such as communication and products such as aeroplanes.
- The educational potential of the general population has been exploited increasingly well over the past 150 years, which has conferred tremendous scientific, technological, social and economic advantages.

Population growth shows just how successful cultural evolution has been. Between 4,000 and 10,000 years ago, the earth was inhabited by an estimated 4 million to 8 million humans. By the time of Christ's birth, this number had risen to around 170 million. Over the next 1,000 years, the world population rose to 300 million people, and after another 800 years it reached a billion. In the last 100 years, it has grown exponentially: from 2 billion in 1930 to 6 billion in 2000, and nearly 8 billion in 2019.

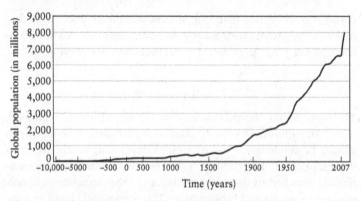

Figure 1.11: The development of the world population, from 10,000 BC to 2007.

Cultural and technological progress has delivered ongoing improvements in quality of life and life expectancy – to the extent that overpopulation and ageing populations are increasingly a cause for concern.

And this progress has come with some significant collateral damage. The exploitation and destruction of the environment have taken on global proportions. Humans live predominantly in the environment they created, having largely alienated themselves from the natural world. The exponential development of cultural progress increasingly overwhelms us, despite our highly developed abilities.

Cultural progress has taken a variety of courses in different geographical regions, depending on the climate, availability of food and the other factors previously mentioned.[33] While the Western world has undergone rapid development over the last 200 years – often to the detriment of other regions – the !Kung people who inhabit the Kalahari Desert of southern Africa are still living as semi-nomadic hunter-gatherers. They are by no means less advanced than Europeans: they too have highly developed cognitive abilities, a nuanced language, and their own world of symbolic ideas and activities. They live according to their own moral and social rules and have their own world of myths and rituals. They know the natural world around them – the habits of animals, for example, and the healing properties of

39

plants – much better than any student of biology does. And they make the leap from pre-Stone Age to postmodern without any problem when they are given access to the necessary education.

Social evolution

The roots of our social behaviour stretch right back through evolution. Mammals have probably been using attachment behaviour to ensure the development and survival of their young for the past 200 million years. Human children spend their first fifteen years unconditionally attached to their parents and any other attachment figures who feed, care for and protect them. The parent–child bond and the family structure retain their vital significance to this day (see Chapters 4, 7 and 10).

The second social structure important for survival is the community or tribe. We have only limited evidence of how our earliest ancestors lived; we have to rely on discoveries made by anthropologists and evolutionary biologists.[34] Scientists work on the assumption that our ancestors led a life similar to that still led (or led until recently) by the !Kung people in the Kalahari Desert, the Yanomani in the Amazon rainforest or the nomadic Nenets of Siberia.

If this is the case, then those early communities consisted of family groups of various sizes, in which different generations and types of relatives lived together. They were made up of between 50 and 500 people, all of whom knew each other from childhood and shared the closest of existential, social and cultural bonds. The members of a tribe spoke the same language and shared myths about their origins and the world around them. They celebrated customs and traditions developed over time. Depending on how densely populated their territory was, contact with other groups might be frequent, occasional or even non-existent.

The archaic community was a closed space within which all social, cultural and commercial activities were carried out and to which, until around 10,000 years ago, there was no alternative. Each member brought their own individual talents and knowledge to the group, and profited from the activities and achievements of others. Children were raised with the support of the whole community. And the way tasks were allocated corresponded to the diversity among individuals. This tight web of mutual reliance was useful in many ways for individuals as well as for the community as a whole. It gave people a sense of intimacy and of existential, social and emotional security. The price for this was the huge pressure to conform placed on the individual by the community. As we know today, the archaic community was no paradise; for some years now, scientists have been debating the degree to which violence ruled both within and between tribes.[35]

Having existed for 200,000 years, these extended families and communities with their close social, economic and cultural ties have become small family units within an anonymous mass society in the space of a few generations. Human needs, such as emotional security and a sense of belonging, have, however, remained the same. We have an extremely strong need for security and social cohesion. We want to feel we belong and have a secure social status in the community. As the structures of family and community have weakened in modern society, emotional insecurity and social isolation have steadily increased. Family and community are still almost essential for survival in our own age (see Chapters 7 and 10).

Modern society has made humans very dependent. In the past, the work of finding food, caring for children, looking after the sick and protecting possessions was done by the members of a small community. In modern society, these things have been taken over by anonymous institutions such as the food industry, nurseries and the police. The absence of familiar people and our constant dealings with strangers cause stress in an increasing number of areas, including the workplace (see Chapter 7). Chapter 10 explores the ways in which we might transform society by focusing on human needs.

FUNDAMENTALS FOR
THE FIT PRINCIPLE

Our history as *Homo sapiens* appears extremely modest when we consider that it covers a mere 200,000 years – a ridiculously small timespan in evolutionary terms. If we were to reduce the 600 million years that have passed since the appearance of multicellular organisms to a 24-hour period, then the 6 million years since our earliest ancestors branched off from primates would take up just 14.5 minutes, and our 200,000 years as *Homo sapiens* would amount to a mere 29 seconds, the blink of an eye in evolutionary history. Still, in this short time humanity has been extremely successful – and this success is due to our highly developed needs and abilities. They largely determine our thoughts, feelings and actions, and make a fundamental contribution to what we regard as the meaning of life.

Each human being is unique

> Every creature can only be good in his own way.
>
> Sophocles (497/6–406/5 BC)

The huge diversity among people is not down to fundamental differences between their characteristics. Everyone has the same set of needs and abilities. It is the different levels at which these are set that make each person a unique being. Basic needs for things like self-development, material security or social recognition are felt more or less acutely by different people. There are artists whose sole desire is to realize their talent as best they can. Other people spend their whole lives striving for material security and amassing a great fortune. And others again – politicians, for instance – live for public esteem and an elevated position in society (see Chapter 4).

Like our needs, the mental and physical abilities we use to help us satisfy them are set at different levels and employed in different ways. For example, people with very well-developed motor skills might become footballers, marathon runners or dancers (see Chapter 5). There is no physical or mental characteristic that is the same in all humans. It is the incredible range of ways in which our many and various abilities are developed and combined that creates this great diversity. And the thing that sets humans apart above all is that they are conscious of their individuality and want to exercise it. Accepting and dealing with our own individuality and the uniqueness of other people is a lifelong challenge.

Our drive to understand the world

The noblest pleasure is the joy of understanding.

Leonardo da Vinci (1452–1519)

In 200,000 years of interactions between *Homo sapiens* and their environment, *Homo sapiens* have consistently been the weaker of the two. But at some point humans began to make use of their cognitive abilities to ward off existential hardship. Their permanent curiosity, their eagerness to learn and their thirst for knowledge were no longer employed only to help them satisfy elementary needs like thirst, hunger or protection from threat (see Chapter 3); they became the biggest drivers of cultural evolution. And since that point humans have never looked back: they have an insatiable desire to understand the world better and better. They continue to penetrate ever further into the micro-world of atoms and the macro-world of the universe. They invest billions in CERN just to find out whether the Higgs boson really exists, and spend more billions sending a mission to Mars just to investigate whether there is water or perhaps even life on the planet. The human thirst for knowledge has given rise to huge scientific, technological and economic progress worldwide.

This thirst drives society – particularly the education system and the economy – and each of us as individuals, too, forcing us into competition on many fronts: better education, greater achievements, smarter children. We have to compete, even though we don't start from a level playing field: there are huge differences between people in terms of both genotype and living conditions.

Our drive to control the environment

We are living in dangerous times. Man controls nature before he has learned to control himself.

Albert Schweitzer (1875–1965)

The desire to keep existential hardship to a minimum didn't just make us want to understand the environment increasingly well: it made us want to control it. For this reason, humans try to make the greatest possible use of their environment, and keep intervening to control it still further. The drive to understand and control the world has intensified over the course of human history, as human adaptation has led to an extraordinary improvement in living conditions. Our need for understanding and control is developed to a level unique among all living things. It spurs us on both as individuals and as a collective, and evolutionary experience has written it firmly into the human psyche. Even small children display the need to build themselves a 'den' and crawl into it.

When we interact with our environment today, we are the stronger player. Anthropologists, geologists and climate researchers speak of a new era: the Anthropocene (see Chapters 7 and 10).[36] If the drive to control our environment once secured our survival, today it threatens to take us in the opposite direction. The exploitation of the environment and man-made environmental disasters threaten the existence of an increasing number of living things – and, in the longer term, human existence as well. We are faced with a great challenge: the urge to tame the natural world was once useful to us, but it has begun to take on a destructive character. We need to learn to deal with our existential anxieties and lower our material expectations. And this will involve taking responsibility for our actions and becoming more humble.

Figure 1.12: (*left*) Lars 'builds' a little house for himself and his sister; (*right*) Ushguli, a village with watchtowers in Georgia, built in the Middle Ages.

Striving for emotional and social security

The gifts of others thou hast not,
While others want what thou hast got;
And from this imperfection springs
The good that social virtue brings.
Christian Fürchtegott Gellert (1715–69)[37]

Over the course of evolution, humans have become deeply social and emotionally needy – qualities that have proved very useful in terms of evolutionary biology, since they hold a small community together and ensure that offspring are raised well. Humans may be equipped with tremendous cognitive abilities, but they are only capable of living alone for a limited time. As the poet Fürchtegott Geller notes, we all have strengths, but our weaknesses mean we can only exist in a community of familiar people.

Our attachment and relationship behaviour, our emotions and the way we become socialized have taken shape over hundreds of thousands of years in family tribes and communities. At every stage of our lives we depend on a reliable, sustainable network of relationships for our mental and physical wellbeing. As children, we develop a strong bond with our parents and other attachment figures, who provide us with security and affection. Once we are adults, we want to live in a dependable community with familiar people. We need a certain measure of emotional security, and we desire social recognition and a position that suits us within the community. We spend a great deal of time and energy attaining these things, by many different means, including maintaining an attractive appearance, performing well in various fields and amassing material wealth. But for several decades our success has been gradually dwindling.

Technological and economic progress increasingly dictates social structures and therefore increasingly controls how people live (see Chapter 7). In the past, the collective use of individual abilities and a sense of solidarity formed the core of a successful survival strategy. The social bonds of family and community that are necessary for this strategy have, however, been getting weaker as anonymous societies have become more widespread. In Chapters 7 and 10 we will look at the following question in detail: are we made for all kinds of social environments, or do we depend on a stable existence surrounded by familiar, reliable people – a social environment that corresponds to our emotional needs and social abilities?

Our efforts to live in harmony with the environment

What is happiness? It is the harmony between a character and his fate.
Thus it can be bestowed by nature, created by the spirit.

Ernst, Freiherr von Feuchtersleben (1806–49)[38]

All living things developed from a close, reciprocal relationship between genotype and environment, a relationship which is billions of years old and exists to this day. Though very simple in terms of function and structure, bacteria have found ways to exist under a huge variety of conditions for more than 3 billion years. Snails can neither see nor hear, but their sense of smell, taste and touch are sufficiently well developed to have survived for the past 500 million years on land and in water. Modern man has many highly developed abilities. You might think we have become largely independent of our environment. But we too must continually adapt to the many and various demands of the environment in order to fulfil our needs as well as possible. This effort to live in harmony with ourselves and our environment is the core of the Fit Principle.

2

The Combined Effect of Genetic Predisposition and Environment

What our genotype accomplishes, our environment cannot achieve – and vice-versa.

If a boy can't speak by the age of two, his parents will worry: is his delayed speech development genetic, or are we not communicating with him enough? They will parent the child differently depending on which assumption they tend towards. If a nine-year-old girl is still having trouble reading, her teacher will assume that she hasn't had enough practice, or perhaps that she is genetically predisposed to develop more slowly than other pupils. Depending on the significance the teacher attributes to genotype on the one hand and encouragement on the other, he might give the girl more homework, send her to remedial lessons, or exercise patience with her and reassure her parents that their daughter will learn to read soon enough. In my consulting room, I have observed on many occasions the extent to which parents and teachers allow their attitude to a child to be influenced by the importance they ascribe to genotype and environment – although they seldom refer to them in these terms.

For adults, too, there is no avoiding the questions we continually ask ourselves: where do my talents lie? Can I achieve more if I make more effort, or am I overstretching myself? Where are my limits? But these considerations aren't just important for the individual; they play a crucial role in society as a whole. The extent to which equality of opportunity is practised in the education system, for example, essentially depends on the attitude taken by education ministers and educators. Do they believe that children from families with low levels of education simply need more encouragement to be successful, or do they see these children as actually less gifted?

47

How much is determined by our genotype, and what does our environment allow us to do? We make assumptions about these things all the time, within our families, schools, society and the economy, and these assumptions have very tangible effects on us and on our fellow humans. If we believe our genotype is the deciding factor, we fall prey to a kind of fatalism. We feel our abilities and behaviour are completely fixed. Our whole lives are predetermined. But if we see ourselves solely as products of our environment, we overestimate its influence and end up with a different kind of fatalism. Destiny is pulling the strings. And if we believe that humans are purely self-determined creatures, dependent on neither their genotype nor their environment, then we overestimate the freedom we have to shape our own lives, and burden ourselves with too much responsibility. Most people probably work on the assumption that genetics and environment both have an important role to play.

In order to understand the combined effect of these two factors, it is helpful to take a look at child development. What predispositions do children inherit from their parents, what does the environment contribute to their development, and what independent contribution do children make themselves? What meaningful support can parents and teachers provide? And what significance do genotype and environment retain into adulthood? I have spent many years searching for answers to these questions, and I hope that the remarks that follow will both provide insights and help in practical dealings with children and adults.

HOW GENOTYPE AND ENVIRONMENT
WORK TOGETHER

There have been many attempts to explain the interplay of genotype and environment over the years. Ideas have ranged from models in which development is determined exclusively by genotype, to those that regard development as largely dependent on the environment. Between these extremes lie various explanatory approaches that see development as the combined effect of genotype and environment.[1] One such model still in frequent use was first put forward by the psychologists Arnold Sameroff and Michael Chandler.[2] The core of their 'transactional model' is the ongoing relationship of interactions between a child and her environment. They see development as a process of reciprocal influencing: the environment has a constant effect on the child, but the child also has an effect on her environment – a kind of endless game of ping-pong. The unsatisfactory thing about this model is that the roles of genotype and environment are not fixed. How a child will develop remains uncertain and therefore unpredictable. In theory it seems possible that a child might develop in any number of directions – an idea that doesn't really correspond to our general experience of life.

In order to comprehend the combined effect of genotype and environment, we must first remind ourselves of the fact that physical and mental characteristics such as height and intellectual capabilities are passed on from parent to child in varying measures. And, secondly, we need an understanding of what environmental factors such as nutrition and education contribute to development.

Parents and children: the same but different

Let us first take a look at the genotype. This is often understood as a person's genetic make-up, the genome. But when we refer to the genotype here, we don't mean the genome; we mean the organism brought into being by the genome in the course of development. The manifestation of all a person's physical and mental characteristics – the phenotype – isn't always the same as the genotype. If a woman smokes a lot during pregnancy, for example, or passes an infection on to the child before birth, the child's development will be affected and her phenotype will differ from the genotype.

Some parents are astonished to realize how different their children are from them – or how similar they are. Other parents marvel at the degree to which abilities and behaviour can vary between siblings, or how similar twins are. In the Zurich Longitudinal Studies, a lot of parents asked: What can we expect from our children? When they grow up, will they be as smart as us, or even smarter?

Height is a good example of how much children can differ from their parents. It's a phenomenon we are all familiar with: tall parents have taller than average children, and short parents have short children. The British scientist Francis Galton took an interest in this relationship as long as 130 years ago.[3] He explored the statistical relationship between the size and weight of a parent generation and their offspring, first looking at peas, and then people. Having measured the size and weight of several thousand peas, and 900 parents and their grown-up children, he discovered a universal law that holds true for most living things, including humans: the further the parents are from the mean value in their own population, the greater the probability that their offspring will tend back towards this mean. The law is known as *regression to the mean*. It applies to all characteristics and talents that are inherited 'multifactorally' – meaning that not one but several genes are involved in forming them, as is the case with height.

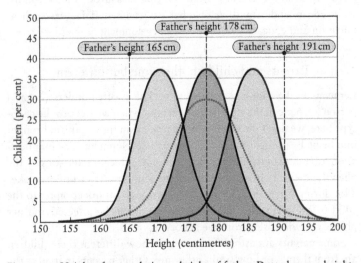

Figure 2.1: Height of sons relative to height of father. Dotted curve: height distribution within the population. Centre curve: height of sons whose fathers are of average height (178 cm); 50 per cent of sons are taller and 50 per cent are shorter than their fathers. Left curve: height of sons whose fathers are 165 cm tall; 84 per cent of sons are taller than their father. Right curve: height of sons whose fathers are 191cm tall; 80 per cent of sons are shorter than their fathers.

If fathers are of average height (mean: 178.5 cm), then their sons are most like them. As adults, 50 per cent of these sons are taller and 50 per cent are shorter than their fathers. But when fathers are just 165 cm tall, 84 per cent of their sons grow up to be taller than their fathers. They can grow as tall as 180 cm. Only 16 per cent are the same height or even shorter than their fathers. And for fathers who are 191 cm tall, the exact opposite is true: 84 per cent of their sons grow up to be shorter than their fathers. A few are merely of average height. Only 16 per cent are as tall as or taller than their fathers. The amount by which these sons' height can vary is always less than the range within the general population (the dotted curve here). A random difference from the father's or mother's height is impossible. If you want to give as accurate a prediction of a child's height as possible, you need to take into account the height of that child's mother and father.

Galton's law also applies to intellectual abilities, which are also shaped by multiple genes. The following graph shows that mothers with an average IQ have daughters whose intellectual ability is distributed around the mean. But where mothers occupy an extreme position on the normal distribution curve, their daughters tend back towards the mean, as with height. As adults, only 16 per cent of daughters whose mothers have an IQ of 130 or above grow up to become their mother's intellectual equal or more intelligent than her; 84 per cent are less intelligent, some of those being only of average intelligence. The law also applies at the other end of the scale: 84 per cent of daughters whose mothers have an IQ of 70 go on to have a higher IQ than their mothers, and only 16 per cent have an IQ of the same or less. And of course, as with height, this law also applies to fathers. For a more accurate prediction of a child's intellectual ability, the IQs of both parents have to be taken into consideration.

When choosing a partner, tall people tend to opt for someone tall, while shorter people choose someone short (see Chapter 1; correlation coefficient around 0.5). This is known as 'assortative' mating, and it often also takes place in relation to social status, level of education and IQ.[4] One might therefore expect that down the generations the range of heights would gradually increase: tall people would get taller, and short people would get shorter. In a similar way, over a few generations intelligent people would get more intelligent, and less

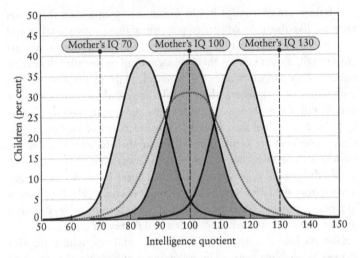

Figure 2.2: IQ of daughters relative to IQ of mothers. Dotted curve: IQ distribution within the population. Centre curve: IQ distribution among daughters whose mothers have an average IQ (100); 50 per cent have a higher IQ than their mothers, and 50 per cent have a lower IQ. Left curve: IQ distribution of daughters whose mothers have an IQ of 70; 84 per cent of daughters have a higher IQ than their mother. Right curve: IQ distribution of daughters whose mothers have an IQ of 130; 84 per cent have a lower IQ than their mother.

intelligent people would get even less intelligent. But, as Francis Galton rightly predicted, that doesn't happen. When living conditions remain the same, the range of both height and intelligence remains very constant down the generations. The same observation can be made for all multifactorally inherited characteristics – and not just in humans, but even in simple organisms like peas.

If both parents are extremely gifted intellectually, it is quite unlikely that these gifts will be passed down to their children at the same level. The famous physicist Albert Einstein and his first wife, Mileva, who met and fell in love at the Zurich Polytechnic Institute (now ETH Zurich) in the final years of the nineteenth century, had three children.[5] Were these children just as clever as their parents? Little is known about Lieserl, the eldest. She was born in Siberia and either died in infancy or was given up for adoption. The second child, Hans Albert, attended a grammar school in Zurich, studied at ETH Zurich like his parents and became a professor of hydrology at the renowned University of California in Berkeley. But his academic legacy was nowhere near as exceptional as his father's. The youngest child, a very sensitive, musically gifted boy called Eduard, developed schizophrenia at the age of twenty. He spent many years in a psychiatric hospital and died when he was sixty-two.[6] Between the sixteenth and nineteenth centuries, the Bach family produced an unusual number of town musicians, organists and composers, but only one of them – Johann Sebastian Bach – was truly exceptional.[7]

Why humans got taller

The genotype is the organic foundation from which a child grows and develops. It determines that child's individual potential for development and the physical characteristics and intellectual capabilities that can be achieved under optimal conditions. But what influence does the environment have? Its contribution to different areas of development, including height, has been investigated in numerous large-scale studies.

In the mid-nineteenth century, compulsory military service was introduced in most countries of Europe, including Switzerland. Since then, the height of all Swiss men has been measured at the age of nineteen, when they begin their national service. This means we have access to data going back 140 years. And it allows us to investigate the way that living conditions, which have improved hugely during that time, have influenced human growth.

In 1875, the average height of Swiss recruits was 163.5 cm, and in 1990 it was 178.5 cm – an increase of 15 cm in 115 years. As the graph below shows, this development plotted a different course in different parts of the country. In urban cantons like Geneva, it began significantly earlier than in rural cantons like Appenzell.

Unfortunately there is no such complete, reliable dataset on height available for women. But based on information from passport registers and mass screenings of schoolchildren, it is possible to tell that over 120 years the average height of women also rose by 15 cm, from 150 cm in the 1870s to 165 cm in the 1990s.[8]

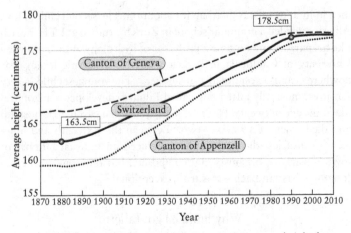

Figure 2.3: Secular trend in Switzerland. Increase in average height from 1875 to 2008 in Switzerland (CH), and in the cantons of Geneva and Appenzell. (source: Staub et al. 2010, 2011)

This continuous increase in height over more than 100 years is what anthropologists call a 'secular' trend.[9] A similar trend has been observed in all European countries, as well as in some countries in Asia and Africa.[10] Numerous environmental factors have contributed to it. In the nineteenth century, the population of Europe suffered from chronic malnourishment and was afflicted by several famines. Vitamin D deficiency in children gave them rickets and restricted their growth. Iodine deficiency led to underactive thyroids, causing stunted growth and goitre. The health of children and adults was affected by life-threatening or crippling infections such as cholera and tuberculosis. Hard physical labour and chronic stress also inhibited growth – and, last but not least, people living in the countryside tended to experience worse conditions than those in cities.

The advent of industrialization gradually started to improve living conditions, particularly when it came to nutrition, levels of education, hygiene and healthcare. People grew taller with each successive generation. Height increased more rapidly at the top of the social scale than among the lower classes. But in the 1990s this secular trend reached all levels of society. And it should be noted that at this point the trend stopped – not just in Switzerland, but in other European countries, too. No further increase in height is to be expected.[11]

The secular trend applied not only to the increase in height, but also to the range of heights within the population. In 1878, the range, defined as the difference between the shortest and tallest men, was 30.8 cm. Their height ranged from 148.1 cm to 178.9 cm (mean: 163.5; standard deviation 7.7 cm). Today, the range is 26 cm, from 165.5 to 191.5 cm (mean 178.5; standard deviation 6.5 cm). This means the range has shrunk by 16 per cent over the past 120 years, the main cause of which is an improvement in living conditions among the lower classes. Most people now get the nutrition they need and are in a good state of health.

But improved living conditions haven't just led to an increase in height; they have significantly increased the rate at which our bodies mature. In the nineteenth century, many people only reached their adult height after the age of twenty. Today, men finish growing at 16–18, and women by 15–17 at the latest. The length of time taken for physical development has decreased by around 4–5 years since the mid-nineteenth century. And this applies to sexual maturity as well as height. Secondary sexual characteristics like pubic hair are appearing earlier and earlier, the length of puberty is decreasing, and sexual maturity is being reached at an ever earlier age. There is particularly good evidence of this acceleration when it comes to the menarche (a girl's first period).[12]

Data from Finland, Norway and the USA shows that in the mid-nineteenth century, girls reached menarche between the ages of seventeen and eighteen. Over the decades that followed, this happened earlier and earlier. Today, girls get their first period at an average age of twelve and a half. But here, too, there is no evidence to suggest that physical maturity will continue to accelerate in the future. The length of time it takes for the human body to develop has decreased substantially over the past 140 years, but the differences in sexual development and height between the sexes have largely remained the same. Today, girls start puberty between eighteen months and two years earlier than boys, and they finish earlier, too. As adults, women are an average of 13 cm shorter than men.

The significance of genotype and environmental factors on growth

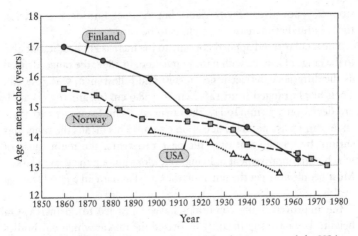

Figure 2.4: Secular trend of menarche in Finland, Norway and the USA (average). The average age of girls at first menstrual period between 1860 and 1975. (source: Marshall et al. 1986)

and physical maturity can be summarized as follows: when living conditions are unfavourable, people are shorter and physical development slows down. If living conditions improve, height increases and physical development accelerates. If living conditions for the entire population are good enough, people stop getting taller and physical development ceases to accelerate. The differences in height and physical maturity between boys and girls remain the same, as long as living conditions are the same for both sexes. These differences are a reflection of different developmental timetables for the two sexes.

The halt in the secular trend also shows that even under good, equal living conditions, people grow to different heights. The differences in height in today's population are largely an expression of individuals' genetic make-up, and not the expression of different living conditions. The population's potential for growth has now been exhausted. It is impossible for an individual to grow taller than the height set by their genotype.

Why humans got more intelligent

The effect of the environment on intelligence is harder to prove than its effect on height. It is much more difficult to take reliable measurements of linguistic and cognitive abilities than it is to measure height, and the environmental factors that influence them are more difficult to determine. And yet we do have a range of studies which, taken together, provide a consistent picture of the combined effect of genotype and environment on intellectual development.

The American political scientist James R. Flynn proved a kind of secular trend for the intelligence quotient, which is now known as the Flynn effect. Flynn studied the development of the intelligence quotient in numerous industrialized countries. He discovered an average increase of 3–5 IQ points per decade between 1940 and 1990.[13]

The Flynn effect is attributed to the following causes: improved nutrition and medical care; fewer children per family, meaning more attention is paid to each individual child; and increased exposure to

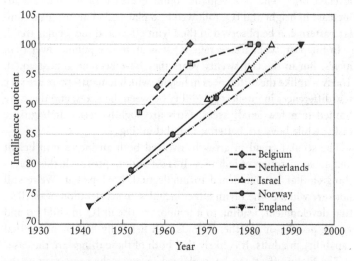

Figure 2.5: The Flynn Effect: average increase in the average intelligence quotient between 1940 and 1990 in selected countries. The IQ values have been re-standardized to a base value of 100. (source: Flynn 1984, 1987)

media. The fact that the increase in IQ since the 1980s was due less to an increase in linguistic ability and more to greater spatial awareness, in tasks like recognizing shapes or fitting together puzzles, speaks for the last of these.[14] The education system made a crucial contribution to the improvement in intellectual ability, as access to education was increased for socially disadvantaged children. In recent decades, the Flynn effect has been barely perceptible in children from the higher echelons of society, though it has still been seen in children from lower down the social scale, and this has led to a significant rise in the average IQ across the whole population.[15] But children from families with low levels of education still haven't completely caught up. The different points at which the Flynn effect first took hold in different countries can be put down to factors such as the state of their economies and education systems, and the proportion of the population belonging to each social class.

Trends reported in Denmark, Germany, France, Britain, Austria and Switzerland show that the increase in IQ scores slowed down dramatically among the middle and upper classes in the 1990s, and eventually levelled off.[16] The Norwegians found evidence of a simultaneous increase in height and IQ, followed by a plateau.[17] This means a similar pattern can be observed in the Flynn effect and the secular trend.

In the past, women's average IQ was up to five points lower than men's. But in the last twenty years they have become almost equal. Today – unlike the differences in height, which remain – there are very few differences in intellectual ability between the sexes, and these are limited to a few small areas. Girls are slightly better at linguistic tasks, while boys are better at visual thinking.

The secular trend in growth involved both an increase in height and an acceleration of maturity. It is difficult to prove whether something similar has happened to intellectual development. We're still not sure whether the Flynn effect includes an acceleration of intellectual development, leading to a temporary rise in IQ in children and young people, or whether it leads to a generally higher intellectual capability in adults. It is likely that both of these things are the case.

The Flynn effect can be explained in much the same way as the secular trend. If living conditions improve, in particular the quality of the education system, then the intellectual capacity of the population

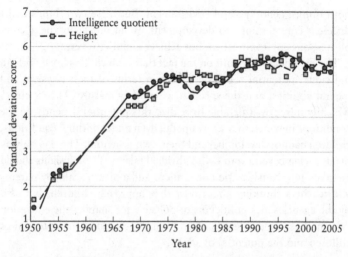

Figure 2.6: Secular trend in height and Flynn Effect on IQ in Norway between 1952 and 2002. Average height and average IQ both increase until 1990, then both parameters plateau out. (source: Sundet et al. 2004)

goes up. If living conditions and education for the whole population are good enough, people are able to realize their individual potential for development. Thereafter, any further increase in ability is not to be expected. The potential for cognitive development cannot be exceeded.

What general conclusions can be drawn from the secular trend and the Flynn effect? The genome is a legacy handed down to us from biological evolution. It forms the organic genotype from which a child is able to develop. It lays down the potential for development and fixes what an individual can achieve under optimal conditions. The genotype alone, however, doesn't prompt any kind of development. This only happens when the environment comes into play, providing nutrition, for example, or learning experiences specific to development. The environment can't generate any characteristics and capabilities, but it can strengthen or weaken them and take them in different directions. Children will learn a language more or less well and speak the language of the social milieu in which they are brought up – German or English, for example. In order for children to develop, then, they need both things: genotype and environment. But these two things make different contributions to development. What our genotype accomplishes, our environment cannot achieve – and vice-versa.

The Fit Principle is built on the fact that each child is born with an individual potential for development in terms of their physical and mental abilities, and the speed at which they mature. The extent to which children will be able to realize this potential depends on the conditions in which they grow up. But even under optimal conditions, children cannot develop beyond their own potential. There is no way of advancing beyond your own individual talents. These insights should be taken into consideration on a micro and a macro scale, by parents and teachers thinking about how they approach children, but also by the agencies and politicians responsible for shaping the education system. The same rule applies to both the development potential of children and the potential of adults.

Different levels of ability

No one would seriously dispute the fact that people are different heights; the variety is too obvious. But when it comes to intellectual and social abilities, which vary even more from one person to another, we are less prepared to accept diversity. When children have trouble reading, for instance, they are urged to make a bit more effort, and are regarded as lazy or stupid. But what goes for height is just as true for cognitive abilities: they are very different for different people.

The PISA studies show the variety of potential talent within a population, and the significance of living conditions – especially the quality of education on offer. In 2009 these studies were carried out in more than sixty countries with very different education systems. The graphs below show the reading levels achieved by students at the age of fifteen in three countries that are at different stages of development.[18] In countries like Kyrgyzstan, where the education system is very poorly developed, 30 per cent of students are completely unable to read (level <1b) and a further 53 per cent only have a very limited ability (1b, 1a); 16 per cent achieve the base level (2); just 1 per cent have a high ability (4, 5); and not a single student has a very high ability (6).

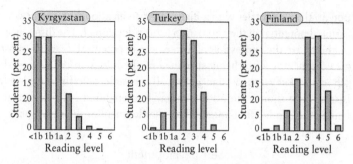

Figure 2.7: Reading ability in Kyrgyzstan, Turkey and Finland. The bars illustrate the number of students (%) who have reached a particular level. Reading levels: < 1b: absent; 1b, 1a: very low; 2: low; 3: moderate, below average; 4: moderate, above average; 5: high; 6: very high. (source: OECD PISA Study 2009)

Once the education system has reached a certain level (as in Turkey), the numbers reading at levels <1b–5 improve. But here, too, no students achieve a level 6. And in countries such as Finland, where the quality of education is high, the level of achievement rises again, especially at levels 4 and 5. But the rise at level 6 is still only minor. It is worth noting that 0.2 per cent of students still cannot read at all (<1b) and 7.9 per cent can only demonstrate a very limited ability to read (1b and 1a).

Even Finland doesn't manage to help all its students achieve a good level of reading: around 8 per cent still can't read, or can only read in a very limited way (<1b, 1b, 1a). And at the other end of the scale those with a high or very high level of ability (5 and 6) still make up just 14.5 per cent of students. This is the case, without exception, for countries such as Canada, Korea and Singapore, where students showed similar high levels of ability in the PISA study.

The results show that, in every country, the population has a very broad spectrum of potential, ranging from little or no ability to read to a high intellectual ability. The PISA studies yield comparable results in mathematics, collaborative problem-solving and science. The better the living conditions and the higher the quality of education within a society, the more people are able to realize their individual potential. In countries with a high-quality education system, the educational potential in the population is fully developed. The differences in ability are the expression of differences in potential. Even the most gifted educator with the most refined methodology in the best of all schools will never succeed in getting all her students to the same level. Instead, she should put all her energy into enabling each student to realize their individual potential for development at their own pace.

Self-development

Genotype and environment both make essential contributions to child development. But there is a third, equally important agent – namely children themselves. Important indications of this are provided by studies in behavioural genetics, most of which have been carried out in the USA.[19] The scientists behind these studies wanted to compare the development of children who are more or less closely related and who grow up in different family situations.

CHILDREN HELP TO PLOT THEIR OWN COURSE

The researchers only accepted children who lived in good family set-ups into their studies. None of the children experienced poverty or emotional neglect, or other conditions that might have had a negative effect on their development. As the table shows, the degree to which the children were intellectually related to their parents and siblings varied widely.

The researchers took intellectual ability as a measure of similarity between the children. A child's developmental quotient (DQ) and intelligence quotient are easier to measure than other psychological attributes such as personality traits. The reliability of DQ and IQ testing is high, but never 100 per cent. If the same person is tested twice with an interval of a few weeks between tests, their results are 80–90 per cent similar (a correlation coefficient of 0.9 to 0.95). Test results vary because the physical and mental condition of candidates fluctuates from day to day, and they may also achieve more or less depending on the time of day the test is done. We must remember that over the course of a child's development DQ and IQ tests look at different abilities. The milestones in motor skills during the first years of life – the age at which children take their first steps, for instance – make a fundamental contribution to the DQ. The IQ doesn't include motor skills. On the other hand, cognitive abilities such as logical and mathematical thinking can only be captured to a very limited extent in DQ tests, while they can be ascertained very well with an IQ test. These limitations should be borne in mind as you read the following remarks.

Levels of similarity between genetic make-up and environment in twins and their siblings who grow up in different family arrangements, and between adopted children and their adoptive or biological parents

	Genetic similarity, %	Social environment
Monozygotic (identical) twins		
Growing up together	100	same
Growing up apart	100	different
Non-identical twins		
Growing up together	50	same
Twins and siblings		
Growing up together	50	similar
Adopted children		
Adoptive parents	0	same
Biological parents	100	different

The graph on page 65 shows a summary of the results from the Louisville Twin Study.[20] This study is particularly revealing because the children were tested repeatedly over the course of their development. A high degree of conformity and a high correlation coefficient point to a similar intellectual ability – between twins, for example. If the level of conformity and the correlation coefficient are low, the levels of ability are very different, for example between twins and their other siblings. The level of conformity and the correlation coefficient tell us nothing about whether the children's intellectual ability is high or low, only how similar their levels of ability are.

The psychologist Ronald Wilson and his colleagues looked at the influence of different degrees of relatedness and family living arrangements:

Monozygotic (identical) twins who grow up together. Of all the children studied, the members of this group are as similar as it is possible to be in terms of their genotype and environment. Their genetic make-up is largely identical and they grow up in a mostly very similar situation. Differences can arise if their foetal development has been different – if one twin has received more nutrients than the other from the placenta, for example – or if their parents have treated them differently.

In the first year of life, the conformity between these children's DQ scores is a mere 50–60 per cent, because there is no way of pinpointing their level of development accurately at this age. But between this point and adolescence the level of conformity rises to nearly 80 per cent and is therefore almost as high as it would be for a single person who was tested twice. This means that the intellectual abilities of identical twins become more and more similar over the course of their childhood. It is impossible to say whether this high degree of conformity is due to their shared genetic profile or their shared environment, or both. But this question can be answered by looking at the ability of identical twins who grow up in different places.

Identical twins who grow up apart. There are a great number of cases in which identical twins grow up apart after being adopted by different families. Although they grow up in different milieus and never meet, their intellectual development proceeds in much the same way as if they had grown up together. At the age of fifteen, the level of conformity is at almost 70 per cent. The psychologist Sandra Scarr interpreted these findings as follows: because monozygotic twins are almost identical genetically speaking, they have the same abilities, interests and likes.[21] They therefore seek out similar experiences in their own family and in their adoptive families, as far as the environment permits. If identical twins grow up apart and both are musically gifted, they frequently play a musical instrument – the same one, more often than mere coincidence might suggest. The twins' personalities and behaviour also influence their natural or adoptive parents

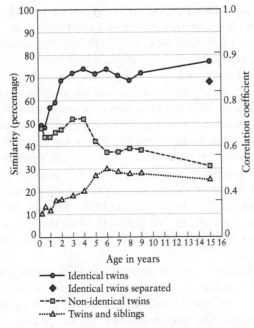

Figure 2.8: Similarity of intellectual ability in identical and non-identical twins and siblings. The similarity of developmental/intelligence quotients is expressed as a percentage on the left-hand side and as a correlation coefficient on the right. The higher the percentage or correlation coefficient, the greater the similarity between the children. There is only a single value available for identical twins who have grown up apart. (source: Wilson 1983, Scarr 1992)

in a similar way, which in turn affects how they treat the child. The theory is borne out by comparing these children with non-identical twins and their siblings.

Non-identical twins who grow up together. These twins share around 50 per cent of their genetic make-up – the same degree of similarity as other siblings have. What makes them special is that they are the same age and therefore grow up at the same time, in the same milieu. In their first year of life, the similarity in their intellectual development is between 40 and 50 per cent. In the years that follow, however, non-identical twins increasingly part ways. They have different interests, make their own friends, belong to different social groups as teenagers and thus have different experiences. By mid-adolescence, the degree of similarity has dropped to a little over 30 per cent.

Twins and siblings who grow up together. These siblings also share 50 per cent of their genetic make-up and live in the same family. But because they are different ages, they experience that family differently, and the similarity in their intellectual development runs in the opposite direction to that of non-identical twins. In the first years of life, this similarity is between 10 and 20 per cent, significantly lower than for non-identical twins. As development progresses, however, twins and their other siblings become increasingly similar. In adolescence, the degree of similarity is almost the same as for non-identical twins. The 50 per cent of their genetic make-up that they share leads them to seek out similar experiences during childhood, albeit at different times.

Further indications of how genotype and environment combine are provided by studies that look at the development of adopted children.[22] The conformity between the child on the one side and the adoptive parents and biological parents on the other is once again measured with the help of DQ and IQ tests. In the first years of life, the correlation coefficient between the children and their adoptive parents is 7 per cent (r 0.27), but between the children and their biological parents it is 15 per cent (r 0.38). Their development is therefore determined a little less by the milieu in which they grow up and a little more by their genotype. Up to adolescence, the conformity between the children and their adoptive parents went back to 0. But the correlation between the young people and their biological parents, with whom they had never lived, was still at 9 per cent (r 0.31). These findings can best be explained by the fact that the adoptive parents were able to determine to some extent what the children experienced in their first years of life. But the older the adopted children became, the more they themselves shaped their environment, by seeking out experiences that fitted their genotype. The decrease in similarity with the adoptive parents doesn't mean the children learned nothing from them. Their potential may come from their biological parents, but the extent to which a child is able to realize that potential is in part determined by the adoptive parents, for example when choosing a school.

If you are a parent, you may well have wondered how much you have in common with your child. Depending on which study you look at, the level of conformity is estimated at between 16 and 25 per cent (r 0.4–0.5) – which may be a slight disappointment to some parents. One important factor in why parents and children are so dissimilar is sexual selection and reproduction; another is Galton's law (regression to the mean, see above). It is no surprise, then, that while not all children are fundamentally different from their parents, many are: they have different interests, and want (and need!) their own experiences. Whatever the level of conformity may be, parents have the hugely important task of creating the conditions in which their children can fulfil their individual potential for development.

Children are active and selective

The results of these studies in behavioural genetics show that children increasingly shape their own development over the course of their childhood. Sandra Scarr put forward a model of development that provides a plausible explanation for the studies' results.[23] The model can be seen to work in everyday home and school environments, and gives important clues as to how we should treat children. It rests on the following assumptions:

Children are active. Children develop themselves. They are led by their own curiosity.

Children want to learn independently. If experiences are forced on children, their motivation to learn will be lost.

Children are selective in their learning behaviour. Children aren't sponges, soaking up everything the environment has to offer. They don't just want any experience, they want experiences that will further their development – starting from the stage they have already reached.

Children's behaviour and personality influence their social surroundings. This behaviour also has an effect on how parents and teachers treat children.

The older children get, the more they determine which experiences they want and what they take away from them. As adults, we often overestimate the influence we can exert over a child. But we have responsibility for something incredibly important: providing a sufficiently varied range of experiences for children to have. We can't shape a child like a lump of clay, but we can provide or withhold experiences specific to that child's development. By shaping their environment in a certain way, we make a crucial contribution to how far children can realize the abilities contained in their genotype.

In the first years of life, the experiences a child is able to have are in large part determined by her family milieu, and most of all by her parents. The older the child gets, the more experiences she wants to choose, both within and, increasingly, outside the family. She is guided by other attachment figures such as teachers, but above all by other children and their surroundings. And when that child becomes a teenager, her experiences are influenced most by her peers.[24] Like children, young people aren't passive but highly selective. They choose friends based on their strengths, their inclinations and needs, and they seek out experiences that fit with their talents and interests. These may be a long way from their parents' talents and interests, or not far removed at all. The Zurich Longitudinal Studies measured the development of hundreds of children, and they show how strongly a child's development is determined at first by her social environment, and then increasingly by the child herself. I have particularly vivid

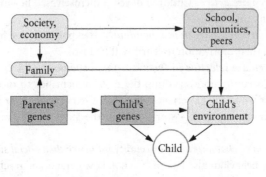

Figure 2.9: The combined effect of genotype and environment in childhood. Dark grey: genotype; light grey: environment; white: child.

memories of identical twins Jakob and Robert. Their parents paid close attention to their linguistic development. Maja, their eldest daughter, had spoken her first words at twelve months and was starting to form sentences at nineteen months. Her parents were proud of their bright little girl. Laura, the second-eldest daughter, had begun to speak at eighteen months, which was at least still in line with average development. But with Jakob and Robert, the parents had to wait until the twins were twenty-seven months old before they said their first words. Three factors had contributed to the delay in their linguistic development. First, speech tends to develop more slowly in boys than it does in girls.[25] The second reason was far more important: the parents, in particular the mother, had very little time for the twins because they also had the two older children to look after. The situation was complicated by the fact that the mother spoke Chinese and the father Swiss German. Multilingualism often leads to a temporary delay in speech development during the first few years of life. The extent to which Robert and Jakob's linguistic development was limited by these three factors in their first two years is shown in the illustration below. In the years that followed, they started to catch up. They communicated with their sisters and the children in their nursery and kindergarten. When they started school, their linguistic abilities were already average, and by the time they were nine their development was actually above average. The more the twins were able to widen the circle of people they spoke to, the better they were able to realize their linguistic potential. Since they had the same genotype and grew up in the same milieu, their linguistic development was very similar.

The process of getting back on to the developmental track laid out by the geno type is a much-studied phenomenon known as 'catch-up growth'.[26] It means that children who didn't get enough nutrients in the womb and were underweight and small at birth more or less catch up on their growth in the first years of their life. As the case of Jakob and Robert shows, accelerated development to compensate for a developmental setback is a phenomenon that can also be seen in areas such as cognition and language.

The boys' progress illustrates a principle of developmental biology: children are not passive creatures shaped by their environment. Nor

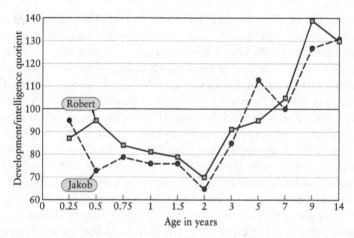

Figure 2.10: Jakob and Robert's linguistic development over the first fourteen years of their lives. The bold horizontal line represents average development. (source: Second Zurich Longitudinal Study)

do they want just any old experiences; they want those necessary for their individual development. Children follow their own developmental path, actively seeking out learning experiences specific to their development – as long as the environment permits it and doesn't force useless experiences on them.

But how do children actually learn? What do we mean by experiences 'specific to development'? And do adults learn in the same way as children or differently? We will look at these questions in Chapter 3.

FUNDAMENTALS FOR
THE FIT PRINCIPLE

We are not marionettes, dangling helplessly from the strings of our DNA and brought to life by Genome the puppet master. The genome certainly gives us our potential. But we are the ones who realize it, by making targeted use of the environment. We are not polymaths: our genotype sets out the abilities we could realize under optimal conditions, and there is no way for us to develop beyond that. If we try, we just make ourselves unhappy. This applies to both children and adults. We find it difficult to accept the limitations of our own potential and that of our fellow humans, in particular our children. This is true on a micro and a macro level, at home and at school, and in society and the economy.

Why it makes sense to fall as well as rise

No one likes going from a horse to a donkey.

Traditional saying

Children can be just as able as their parents, or they can be more or less able than them. This means that children can fulfil or exceed the expectations their parents place on them – but they can also disappoint them. We must also recognize that society and the world of work place different values on different talents. Parents pay particular attention to areas like language and mathematics, because they think these things will be especially important for their children when they grow up. If children have a musical talent, their parents will be pleased, but they will also place much less importance on it.

Parents' concerns have a lot to do with their expectations for their children. They hope their children's lives will be just as good as their own, if not better. But, much to their disappointment, this hope isn't always fulfilled. According to the law of regression to the mean, the more able the parents, the more likely it is that their children will be less able than them. But conversely: the less able the parents, the more likely it is that their children will be more able than them. In every generation, some people rise, and some fall. Parents are understandably very reluctant to hear that their children might fall, and the idea can cause great anxiety. The following illustration describes the statistical probability of rising and falling in social and professional terms in Switzerland.[27]

It is more common than we might think for children whose parents are academics or hold higher managerial positions not to reach the same high professional status: 43 per cent of these children 'fall', meaning that as adults they have a lower status than their parents; 28 per cent go on to become low-level managers or skilled workers,

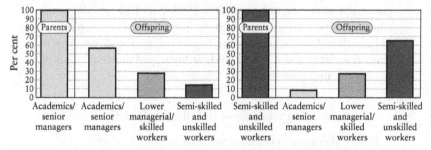

Figure 2.11: Socio-professional rise and fall between parent and child generations. The offspring were between forty and fifty years old at the time of the survey: (*left*) parents working as academics and managers; (*right*) parents working in semi-skilled or unskilled jobs. The bars show the percentage of children rising or falling into each of the three job categories. Only 57 per cent of offspring whose parents worked in academic jobs and in management were working at the same level. (N = 485) (source: Levy et al. 1997)

and 15 per cent become unskilled workers. At the other end of the scale, 27 per cent of children whose parents are unskilled workers become low-level managers and more highly qualified workers, and 8 per cent even become academics and top managers. Whether a society allows people to climb the status ladder depends on the freedom to choose that exists within the education system. Equality of opportunity in schools and access to higher education are the deciding factors for a rise in social status, alongside family and cultural influences.[28]

In our society, descending the social scale is unjustly regarded as failure. A fall can also make sense: it protects people from being over-stretched, both as children in school and as adults in the workplace. In the long term, a fall saves people from choosing the wrong career, being constantly overstretched and inevitably failing. Taking a lower-status option means their sense of self-worth and self-efficacy remains intact. Just because you don't achieve the same social status as the family you come from, you shouldn't regard your life as a failure.

Some parents try to realize their own ambitions through their children, and in so doing stop their children from realizing themselves[29] – an uncomfortable truth, but one we should keep in mind. There are those parents who will do everything in their power to prevent their less talented children from 'falling'. They believe an academic career befitting their social status will materialize if only they put enough pressure on their children. And this can end very badly, if parents put their children in situations where they are hopelessly overstretched. Pressure doesn't guarantee a career – whether it comes from the parents, or the child because she is desperate to fulfil her parents' expectations. Burnout doesn't affect just adults; young people and even children are increasingly affected by it.[30] Parents and children will only find a way out of this agonizing situation if parents change their behaviour and stop seeing low academic achievement or a less demanding career path as failure. If they don't, their children will regard themselves as losers in the short or the long term.

Figure 2.12: If our parents have this many books, then they must be a good thing. A living room full of books encourages children to read, though it doesn't necessarily turn them into bookworms.

Descending the social scale is also in society's interests, because it means that fewer people rise into positions for which they're not suited. There are always far-reaching negative consequences when additional support, privileges and networks help people to reach positions in society and business where their incompetence and bad decisions cause great harm. It is in society's own interests to ensure that people have opportunities to rise, but aren't prevented from falling, either – by means of nepotism for instance, where someone's son is brought in to fill a senior position without being sufficiently qualified for it. These developmental errors often begin with school-age children. Less able students from well-educated families are frequently kept at more academic schools with the help of additional tutoring, or sent to boarding schools where they receive special support. Conversely, intelligent students from deprived families with low levels of education, who would be of great use to society and the economy, are prevented from rising up the social scale. An education system that incorporated the greatest possible fairness and freedom to choose would serve not only the individual, but society as a whole.

For children, parents and society, it is best in the long term if all children are able to take a path through school and work that suits their abilities and builds on their strengths. Some parents may find it difficult to swallow, but it is the only way to be fair to their child: such an attitude accepts human nature and respects children for who they are.

A society for all talents

Bear in mind that the diversity of humans might be their greatest strength.

Anon.

Just as parents and teachers must learn to deal with the different levels of ability among children, society must also find ways and means of doing the greatest possible justice to the diverse talents in the population.

Countries like Kyrgyzstan and (to a lesser extent) Turkey are not doing a good enough job of exploiting the potential talent in their populations. They need to make every possible effort to improve their education systems, though this in turn can succeed only when reforms also take place in a country's society and economy. Countries such as Finland have largely exhausted the potential talent in their population. They face a challenge of quite a different kind: they need to do justice to the diversity of talent within the population. The range stretches from people who have never managed to learn to read, to bookworms who devour several books a week. There is a similarly wide spectrum for all other cognitive abilities, including maths and logical thinking (see Chapter 5).

An education system with a duty to provide equality of opportunity doesn't just offer a high level of education to a small intellectual elite: it serves the needs of as many people as possible by taking all levels of ability into account. In an education system of this kind, students aren't all expected to get equally good results and, in the long term, gain equally high levels of competence. A truly fair education system helps all children, taking into account their individual talents, to achieve the highest academic success they are capable of. True equality of opportunity should exist for adults as well as children. It allows

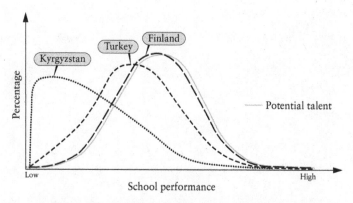

Figure 2.13: Realization of potential depends on education. Light grey curve: potential talents within the population. Kyrgyzstan: realization of potential under poor living conditions and with a very poorly developed education system. Turkey: realization of potential under good living conditions, with a moderately well-developed education system. Finland: realization of potential under optimal living conditions and with a very well-developed education system.

them to utilize their potential and perform at a level in line with their abilities. This of course assumes that the diversity among individual abilities and talents is respected, and people's limitations are therefore also recognized, in society and industry as well as within education. This is something that is happening less and less in today's highly competitive society, and in an economy focused on profit. There is a fundamental conflict here, which necessitates reforms not only in the education system, but across the whole of society and the economy (see Chapter 10).

3
Developing into Individuals

Every human wants to develop their talents,
to become more and more themselves.

Have you ever seen a sparrow flying through dense woodland hit its head on a branch? A sparrow's brain is tiny, weighing less than a gram, but it manages to take in the complex form of a tree in a fraction of a second and steer the bird's flight so that it makes it safely through the tangle of leaves and branches. After an incubation period of just fourteen days, and sixteen days as a nestling, the sparrow's brain is mature enough for it to venture out on its first flight. If its brain were not absolutely prepared for this task it would fall at the first attempt, flutter about helplessly on the ground and in the worst case scenario get eaten by a cat. But the sparrow still has a few things to learn from its environment. How to fly in wind and bad weather, for instance; which trees and bushes grow in its territory, and, most importantly, where to find seeds and small insects to eat.

The human brain weighs in at 1,400 g, as opposed to just one. Unlike a sparrow's brain, it takes around fifteen years rather than just a couple of weeks to mature into an extremely capable organ. It also requires a great many more development-specific learning experiences than a sparrow's brain. Over millions of years of evolution, its highly complex structures and sophisticated functions have been shaped by the combined effect of countless (though not random) experiences. Our brain development has taken quite a different direction from that of the sparrow: we might occasionally hit our heads on a low beam, but we are also capable of thinking about that beam's structures and functions in a way no other creature is.

In this section, we will look at what makes the human brain unique.

What are its core elements, what do we mean by brain maturity, and what do experiences contribute to the brain's development? We also need to understand what real learning looks like: the role of motivation and curiosity, why self-directed learning is so important, and what type of learning has a lasting effect. The answers to these questions will give us a better understanding of children's – and adults' – potential for development, and where they come up against the limits of this potential.

OUR BRAINS

Brain, n. An apparatus with which we think that we think.
Ambrose Bierce (1842–1914)[1]

It sounds like a paradox: the fundamental element that enables the brain to perform very complex tasks is just one type of cell, billions of times over: the nerve cell or neuron. Scientists have found evidence of simple neurons in jellyfish, polyps and worms that existed more than 400 million years ago. And we can say with no exaggeration that without neurons evolution as we know it would never have happened. The things that make the neuron so special are its shape, its length and, most importantly, its ability to communicate. Like all other human cells, it consists of a cell body with a nucleus, and it also has a kind of tail called a nerve fibre or axon, which is often very long. An axon can reach from your spinal cord to your big toe, and its job is to pass on electric signals. In addition to the axon, a neuron has other offshoots called dendrites. They receive signals passed to the neuron by the axons of other neurons. The contact points between axon and dendrites are called synapses. Electrical signals are passed between the synapses by chemical messengers known as neurotransmitters. Neurons can communicate with each other by sending out electrical signals through their axons and receiving them via dendrites. In the brain, neurons are embedded in a supporting tissue of 'glial' cells.

The human brain consists of between 20 million and 100 billion neurons. But this immense number of neurons alone doesn't explain

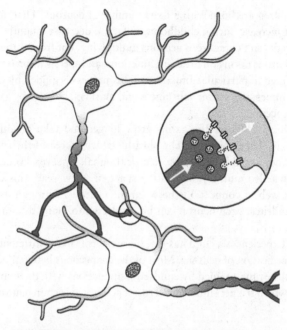

Figure 3.1: The structure of a neuron. White: cell body; dark grey: axons; light grey: dendrites; circle: axon and dendrite synapses. [Enlarged detail: exchange of neurotransmitters in the synaptic cleft.]

its extraordinary capabilities. It is the elongated, hierarchically structured neural networks, developed over the course of evolution and built from scratch by every child as it matures, that make the brain capable of such impressive achievements. After nine months of pregnancy, the brain is already well developed. But it takes another eighteen years to reach maturity, which only happens during adolescence.

How the brain matures

How should we imagine the brain maturing? An obvious thought would be that the brain gets bigger and heavier – growing from 400 g at birth to 1,400 g in adults – because the neurons are multiplying.

But children are born with a fixed number of neurons. That number doesn't increase during childhood; in fact, it decreases slightly. Brain growth doesn't come from neurons multiplying, but from differentiation (neurons taking on different functions) and specification (neurons being fixed in particular functions). This process is guided by genetic programmes, messenger substances and markers, but also by a child's experiences.[2]

During childhood the axons grow longer and take on different functions. They gain a 'myelin sheath', which increases the speed at which they are able to pass on electrical signals by several times. The dendrites also multiply, and in the space of a few years this creates a thick web of connections between the neurons. Through its axon and dendrites, each neuron can be connected to many thousands of others.

The experiences a child has play a crucial role in the differentiation and specification of neurons. Most of the connections between neurons are made or maintained by children's interactions with their environment, which trigger similar patterns of activity in certain neurons – in

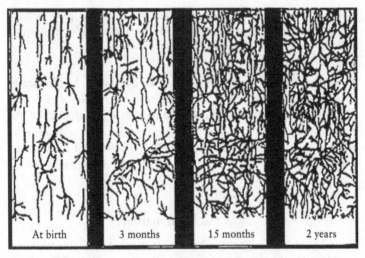

| At birth | 3 months | 15 months | 2 years |

Figure 3.2: Neurons maturing and forming networks in the first two years of life.

the visual cortex, for example. This is where Hebb's rule comes in: 'Neurons that fire together wire together.'[3] Connections between neurons that are active at the same time are maintained and strengthened; connections that aren't activated disappear.

The majority of the synapses are formed during pregnancy and – depending on the region of the brain they are in – reach a peak when a child is between nine and thirty-six months old. After that, the number of synapses decreases, and by the time the child reaches puberty it has dropped by up to 20 or 30 per cent. In other words, an excessive number of synapses is created during pregnancy and early childhood. Which synapses are retained depends on the experiences the child is able to have. The process can be summarized as 'use it or lose it'. Synapses that are used and activated are maintained and strengthened, while those that are not used and remain inactive are destroyed.

We can now see why experiences are so important for brain development: they build and reinforce neural networks. The number of synaptic contacts that each individual nerve cell makes with others rises from 2,500 to 20,000 in early childhood, and certain nerve cells make as many as 100,000 contacts. This process takes place at different ages for different functions and regions of the brain, which also explains why there are 'sensitive' phases in child development. In these phases, which can vary greatly in length, learning experiences are particularly influential, and children are particularly receptive to them.

During childhood, then, brain structures are built up, reconfigured and destroyed. This 'neuroplasticity' starts at a very high level in early pregnancy and decreases steadily after this point. Once a child has been born, nerve pathways and areas of the brain can be restructured only to a limited extent, and for the most part not at all. If the motor cortex, for instance, is damaged in the first years of life by being starved of oxygen, nerve pathways can still be rerouted; later on, neurons and their axonal networks can be built up and modified only in a very limited way. But the plasticity of the synapses and the ability to learn are maintained throughout childhood and – to a decreasing extent – right through to old age.

The web of nerve pathways in the human brain, which can be seen with the help of magnetic resonance imaging (MRI), is a beautiful,

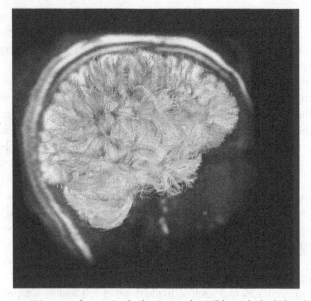

Figure 3.3: Nerve pathways in the brain, made visible with the help of a Magnetic Resonance Imaging (MRI) scan. (source: Martinos Center for Biomedical Imaging at MGH, Boston)

almost awe-inspiring sight. It isn't just the structure of the human brain that is impressive; our brains also have a unique level of capacity that no other living thing possesses. Scientists estimate that the brain of an adult human contains between 20 billion and 100 billion neurons, and the total length of their axons is between 150,000 and 300,000 kilometres – four to seven times the circumference of the earth. Each neuron exchanges signals with 5,000–20,000 others, and for some the number is much higher. The total number of synapses in the brain is estimated at 100 billion to 1,000 billion. John Eccles, an Australian physiologist and Nobel laureate, calculated that the number of possible contacts between neurons was even greater than the total number of atoms in the universe.[4] But what makes the human brain truly unique is the way all these neurons are networked into countless hierarchically organized, functional units, and the way these come together as a coherent whole. Our understanding of this miracle is still in its infancy.

WHAT MATURITY AND EXPERIENCE CONTRIBUTE TO BRAIN DEVELOPMENT

A newborn baby comes into the world with a highly structured brain. But what this brain can achieve is still very limited. In order to develop things like sight and motor skills, the structures in the brain must be made functional, through a process of maturation that involves development-specific experiences. Development in early childhood makes it very easy to see how brain structures mature, and to see which specific experiences a child needs in order to develop abilities such as recognizing faces and gripping objects, or acquiring skills such as writing.

Experiences activate and network our brains

At birth, babies already have certain behaviours developed during pregnancy. A large number of reflexes ensure their survival in the first weeks and months of life. When a baby's cheek touches her mother's nipple, the 'rooting reflex' is triggered and she turns her mouth towards the nipple. And when her mouth touches the nipple, the sucking reflex kicks in: she will latch on and start to suck. When the milk runs down her throat, it triggers the swallowing reflex and she swallows it. When the mother carries the child around, the gripping reflex and clinging or 'Moro' reflex help the child to keep hold of her.

Once a child comes into the world, she craves experiences. She looks at her immediate surroundings and shows a particular interest in the human face.[5] To start with, she can see only the outlines of a face. After a few weeks, she will start to focus more attention on the eyes and eyebrows. And a few weeks after that, she will also look at the mouth. By around four months old a child can understand the face as a whole, and will begin to differentiate between familiar and unfamiliar faces. A few weeks later, she will also be able to recognize expressions. She will now only smile at friendly faces. She will refuse to smile at a neutral or unwelcoming face, and may even start to cry.[6] How should we picture the process that lies behind this development?

The illustration on the following page gives a schematic representation of how hierarchically organized networks are activated by experiences during the first twelve months of life, and how they join together to perform increasingly complex tasks. Visual impressions are carried from the sense cells in the eye via various intermediate stations to the neurons in the visual cortex. These neurons can only recognize the simplest structures: vertical, horizontal or slanting lines. Over the first few months of life, this network is joined hierarchically to larger and larger networks. The increase in the number of neurons being activated allows the child to perceive facial features such as eyes and mouths, then whole faces, and eventually to recognize individual faces and their expressions.[7] This point is not reached by strengthening the neurons themselves, but by joining them together in increasingly complex networks.

In the 1960s, the neurologists David Hubel and Torsten Wiesel showed just how precisely neurons are geared to each other, using the example of stereoscopic vision.[8] Their findings came from studies using young cats, but they also apply to humans. Stereoscopic vision allows us to see objects three-dimensionally and to judge how far away the objects in our immediate surroundings are. In stereoscopic vision, the slightly different visual impressions that each eye gets of the same object are brought together very precisely in the visual cortex. The connections are made in the first years of life, when neurons are activated together by visual stimuli. The neurophysiologist Wolf Singer talks about a synchronization of neuronal activity.[9] Our sophisticated visual perception depends partly on hierarchically organized neural networks and partly on the way these networks are activated and reinforced by visual experiences. If someone misses out on these experiences in early childhood – perhaps due to a visual impairment – then the networks won't be activated and reinforced sufficiently, and this can mean that neural structures don't develop very much, or may even deteriorate. The child will only develop a limited ability to see, or may not be able to see at all.

The development of fine motor skills is a particularly good illustration of how neural networks are extended during childhood. As part

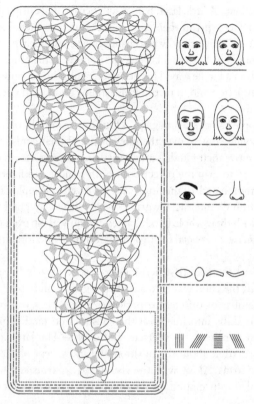

Figure 3.4: The development of the visual perception of faces. Hierarchically organized networks in the visual cortex are activated and connected to each other.

of the Zurich Longitudinal Studies, we conducted a thorough investigation of how children begin to use their hands functionally. In ultrasound scans during pregnancy, we were able to see that children use motor skills long before birth. In the fourth month of pregnancy, the unborn child regularly puts its fingers in its mouth and sucks them, getting to know its own hand. The mouth serves as the first organ of perception. It comes as no surprise, then, that at birth babies have already mastered this behaviour to some degree. They feel their fingers with their lips and tongue. They discover what it feels like when they move them. A little later, babies will occasionally bring their hand up to their face, spread their fingers and watch as they slowly move them around. They are now also using their eyes to learn about their hands. Eventually, babies will touch one hand with the other, exploring their hands in a tactile way. This behaviour links the networks involved in motor function and in tactile-kinaesthetic and visual perception to each other.

By the time babies start to grip, at four or five months old, they have gained enough control over the movements of their arms, hands and fingers to move their hands – albeit very uncertainly – to an object. At first they try to grip the object with both hands, and later with just one. The thumb and all the fingers are involved in gripping (a 'palmar grip'). Within six months, the grip matures: instead of grabbing things clumsily with their whole hand, babies can now grasp even very small objects precisely between the tips of their thumb and forefinger (a pincer grip).

Each time the neural structures for a type of grip have matured sufficiently, the child wants to adopt that grip. Children need to have a large number of different experiences before they can bring their fine motor skills into line with their visual and tactile-kinaesthetic perception (sensitivity to surfaces and depth). The latter gives children information about the position and movement of their fingers, hands and arms. Along with the visual cortex, arms and hands provide detailed information on the object to be grasped. How large is

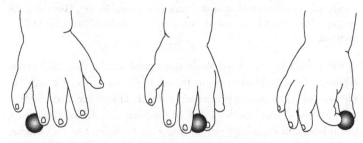

Figure 3.5: The development of grasping behaviour in the first year of life. Palmar grip (using the whole hand) on the side of the little finger at 4 to 6 months (left); on the thumb side at 6 to 8 months (centre); and a pincer grip at 9 to 12 months (right).

the ball, what shape is it, is it hard or soft, and how far away is it? If a child has difficulty experiencing these things – as the result of a visual impairment, for instance – then the ability to grasp objects is delayed, or in the worst cases simply doesn't develop. The neurons responsible for gripping are activated insufficiently or not at all, and cannot work together.

The child needs no guidance for this learning process. It would never occur to parents to demonstrate a pincer grip to their child, or encourage the child to practise it. But learning a motor skill and adapting it to various conditions in the environment is not something that happens at random, either. Children want the development-specific experiences they need in order to acquire an ability like the pincer grip. They pick bread crumbs and other small objects up off the floor. Their behaviour is guided by networks that only respond to specific stimuli. Children also use development-specific experiences to acquire abilities in the other areas of development.

In the years that follow, children learn skills such as writing and drawing. Unlike *abilities* (grasping, for instance), *skills* are not innate. The child acquires skills by learning to use various abilities in combination to perform an activity. Drawing brings together fine motor abilities and ideas about shape and space. Before they can write a story, children need to have mastered the formal elements of language like grammar and syntax and the meaning of words (semantics), as well as the ability to use a pencil. For semantics, children need access to other competencies such as an understanding of numbers and an idea of time (see Chapter 5). For this, they rely on the support of their parents, who will ideally pass on the necessary knowledge and abilities in such a way that their child can master the skill independently. When various abilities are brought together in a skill such as writing, they are saved in hierarchically structured networks as a pattern of neurons that are activated together.

Brain development can be summarized as follows: the structures of the brain, which mature during childhood, create the conditions necessary for development. Development-specific experiences are also required to activate the existing neural networks, to connect them up and adapt them to the environment. At the heart of this process of reinforcing and expanding neural networks is the phenomenon

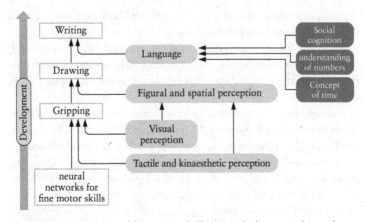

Figure 3.6: Acquisition of fine motor skills through the networking of motor function with other competencies. Fine motor function is first networked with tactile-kinaesthetic and visual perception. Then, further connections are made to other competencies such as figural and spatial thinking, language and the understanding of numbers.

that we call learning. And the need to learn is expressed in children's curiosity and motivation.

But how do we learn as adults? Our basic abilities don't develop beyond our teenage years, as we can see from the example of sequential finger movements in the illustration below. The speed at which these finger movements can be performed increases up to the age of eighteen. After that, it remains largely constant until the age of around fifty. What is true of motor functions also goes for intellectual abilities – which may come as a surprise to some readers. Prior to adolescence, a person's intelligence quotient rises steadily. During adolescence this increase gradually slows down, reaching a peak at the age of about eighteen. It then plateaus out, and remains constant until the age of about fifty. After that, intellectual abilities start to decline – at a rate that varies from one individual to another.

Some adults envy young people the dexterity with which their fingers zip over the keyboard of a smartphone while typing out a message, or their quick grasp of what they are reading. But adults can still learn to use a smartphone or understand books, even at an advanced age.

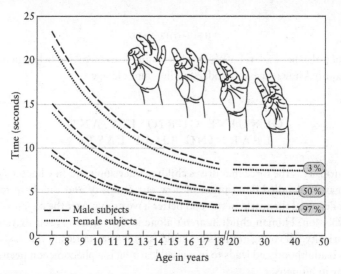

Figure 3.7: The speed of sequential finger movements between the ages of seven and fifty. The thumb touches each finger from forefinger to little finger, one after the other. The test measures the time taken by each participant to perform five repetitions. Male subjects: dashed line; female subjects: dotted line. Fifty per cent denotes the mean; 3 per cent of participants were above 3 per cent or below 97 per cent respectively. (source: Zurich Longitudinal Studies; Largo et al. 2001)

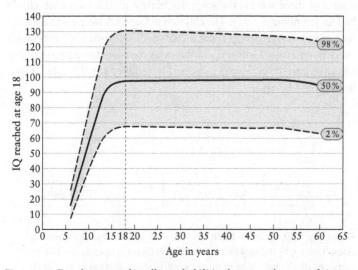

Figure 3.8: Development of intellectual abilities between the ages of six and sixty. The intelligence quotient (measured at various ages in relation to an IQ of 100) is fixed at the age of eighteen. (source: modified from Asendorpf 2005, Baltes et al. 2001)

They also have a level of ability that is unique to them and helps them acquire knowledge and understanding into old age.

GENUINE CURIOSITY AND LEARNING THAT LASTS

Curiosity is the force that drives children's development. In Chapter 1, I gave a detailed description of how our curiosity and the drive to understand the world better and better have grown over the course of evolution. Human children aren't alone in their genuine, self-driven curiosity; all young animals are curious. However, curiosity that lasts into adulthood and leads to increased learning is a phenomenon peculiar to humans.

A lifetime of curiosity

We are born curious. In the 1970s, the developmental psychologist John Watson carried out an experiment with eight-week-old babies that gave a very clear picture of how childhood curiosity works.[10] Over the course of three weeks, the scientists hung a mobile over each child's bed for ten minutes a day. The children in Group A were given a mobile that didn't move; Group B got a mobile that turned for five seconds every minute, and Group C's mobile was connected to a pressure sensor sewn into the children's pillows. The mobile turned whenever they moved their heads.

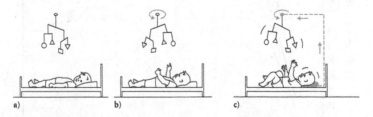

Figure 3.9: Motivation to learn in eight-week-old babies. a) the mobile doesn't move; b) the mobile turns at regular intervals; c) the mobile is activated by the baby's head movements. (source: Watson et al. 1972)

After three weeks, the babies in the three groups were displaying very different behaviours. While the frequency of head movements in Groups A and B hadn't changed, in Group C they had increased significantly. After a few days, the babies in Groups A and B had lost interest in their mobiles, while those in Group C were more interested in the mobile with every day that passed. They quickly learned that they could influence its movement by turning their heads. They found they were able to have an effect on their environment – and became more active as a result. They chattered and laughed more, and displayed more lively facial expressions than the children in the other two groups. Curiosity, then, is aroused and maintained for the longest period when children are able to interact independently with their environments – even in the first weeks of life.

At nine months old, children spend all day crawling round the house, and at eighteen months they are running around the garden. Motor functions are a very good example of how genuine curiosity drives children to acquire abilities through experience. Children take a natural delight in movement, just as kittens chase balls of wool, or foals and calves leap about in the fields. Motor activity is the expression of a natural, irrepressible urge to experience movement. Adults love to see young animals romping about, but parents and teachers are much less pleased by active children. To their minds, this natural urge to move around looks like restlessness, which taxes their parenting skills and disrupts lessons.

Motor activity increases rapidly in the first years of life and reaches a peak in primary-school-age children, declining again as children approach puberty. They are at their most lively and active between the ages of six and ten – the years when they are constantly being told to sit still at school. At every age, boys are slightly more physically active than girls. But the difference between individuals is much greater than the average difference between the sexes. Boys and girls with a high activity drive are around three times more active than those who don't move very much. In adults, physical activity is much lower than in childhood; but a high level of variability remains. There are sixty-year-olds who only move between the kitchen, living room and bedroom, while a few eighty-year-olds take on the challenge of running a marathon.

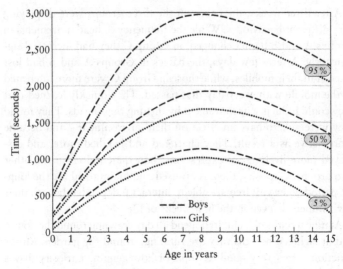

Figure 3.10: Motor activity from the ages of one to fifteen. The frequency of arm and leg movements is measured objectively using actometers attached to all four limbs. Dotted curves: girls; dashed curves: boys. Fifty per cent is the mean value; 5 per cent of children are above 95 per cent and below 5 per cent respectively. (source: adapted from Eaton 2001)

Why do children need to move around so much? One important reason is that they can only develop motor abilities and skills through experience. The other is that children are always having to adapt their motor function to their growing bodies. Over the course of eighteen years the nervous system matures, muscles and bones grow, height and weight increase, and the proportions of children's limbs and torsos change constantly. They need to keep recalibrating their motor functions, aligning them with the sensory impressions they receive from their eyes, their sense of balance, and their muscles and joints. If, for example, a child runs across a field dotted with humps and hollows, he has to keep his body balanced at every step so that he doesn't fall over. If he wants to pick a flower or break off a hazel twig, he has to adapt his fine motor skills to the physical properties of each object. Children want to have experiences in order to acquire other skills as well – writing and drawing, for instance – and to keep adapting these skills to their level of development.

Just as the innate urge to move prompts children to acquire motor abilities and skills through experience, and to adapt these as they grow, curiosity and the motivation to learn trigger the same behaviour in other areas of development. Children acquire abilities and skills in areas like maths, and extend them as their development progresses. There are three basic learning strategies that children use as they develop:

In *object-oriented learning*, children engage independently with the objects in their physical environment. A baby explores a music box with her fingers and discovers that music comes out when she pulls the cord. Constant engagement with objects allows children to improve their abilities, and to keep acquiring skills and making discoveries. For this to happen, they need objects to play with that suit their level of development and therefore arouse their interest.

In *imitative or social learning*, children observe and mimic the people around them. Young children see their parents and siblings using a knife, fork and spoon at mealtimes and acquire the ability to use cutlery through imitation. In the years that follow, children grow more and more interested in what the people around them – particularly other children – are doing. Children are reliant on role models for this type of learning.

Learning through instruction is not really about teaching a child to do something. Instead, adults should arouse a child's curiosity by arranging his surroundings in such a way that he can take the next step in his development as independently as possible.

Adults also use these learning strategies. They engage with objects, such as the functions of a smartphone; they learn by watching other people's behaviour and adopting their value judgements; and they allow themselves to be instructed by more experienced people in areas such as computing.

Every child wants to learn in their own way

Children don't develop uniformly. They all develop at their own pace and have their own individual behaviour when it comes to curiosity.

Figure 3.11: Early developer: three-year-old Lars teaching himself to read.

The motivation to learn varies from one child to another – take reading, for example. Lars, whose development I tracked over many years, wanted to learn to read when he was just a toddler, while other children are only ready to read after they have started school.

The graph below takes the example of three boys to show how different children can be in their motivation to learn over time. Eldar starts to show an interest in letters between the ages of six and seven. At sixteen, his reading competency is fully developed. During his school career his development is average. Lars can read well at the age of three. His reading competency is significantly higher at the age of sixteen than Eldar's – and higher than that of most adults, too. Patrick, meanwhile, doesn't manage to read until he is ten. His reading competency at sixteen is low and remains at this level over the years that follow.

The boys begin to take a genuine interest in letters at different ages, and their willingness to learn also differs at every age. Different levels of curiosity and motivation to learn are often seen as a kind of personality trait, and are also often connected to external factors such as praise and criticism. But above all they are dependent on the harmony between an individual's level of development and the demands of their environment.

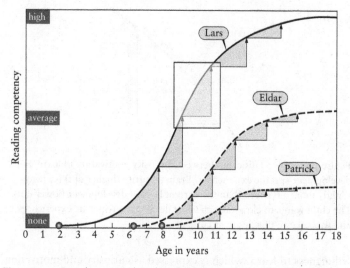

Figure 3.12: Development of reading competency and willingness to learn in three boys. Grey: current developmental level (horizontal line); willingness to learn (vertical line). Note the large differences in willingness to learn and the increases in reading competency between the three boys. (Rectangle: see next graph.)

When children are satisfying their curiosity, they experience a state of 'flow'.[11] Ideally, they will be completely absorbed in their activity and feel a deep sense of contentment. When they first learn to read, children are very proud of being able to recognize letters, and then whole words. In the months and years that follow, the mechanical process of reading becomes less and less important, and children are increasingly interested in what the text says. For Lars, even at the age of four or five, it is the stories that make reading enjoyable. But Patrick faces the formal challenge of reading – using the alphabet to sound out words – for a period of several years.

Children aren't constantly curious and willing to learn, because their brains mature in bursts rather than steadily over time. With each burst, a gap opens up between the current level of development and the slightly more advanced maturity of the brain. This gap arouses the

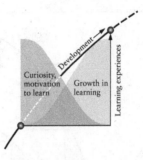

Figure 3.13: The relationship between curiosity, motivation to learn and development-specific experiences. Brain maturity (higher dot) is always slightly more advanced than the current level of development (lower dot). The child wants to close the gap using development-specific experiences, and we perceive this desire as curiosity.

willingness to learn, which is expressed as curiosity and motivation. You will notice an increase like this if a child is ill and has to spend a long time idle in bed. Once he has recovered, he will feel the urge to make up for the learning experiences he has missed and display a high level of curiosity and motivation.

Children's willingness to learn is greatest when a new stage of maturity has been reached. As they have development-specific experiences and raise their level of development, their curiosity and motivation (and, with them, the state of 'flow') will decline. These curves of growth in learning and decline in curiosity form a mirror image of each other. When a child's learning catches up with his current level of brain maturity, he reaches the limits of his resources and his curiosity and motivation are extinguished. Additional efforts, such as parents and teachers often demand from a child, won't lead to any further development. You know that children have reached their limits when their rate of learning starts to slow down and eventually stops altogether. They become increasingly unenthusiastic and finally lose all motivation. Curiosity and willingness to learn return spontaneously when their brain has reached the next stage of maturity.

Some parents and teachers try to provide children with additional motivation to learn using social and material stimuli like praise and rewards, or the pressure of exams and grades. All children need encouragement, but we should praise the effort they have put in rather than the things they have achieved. Adults can support children by giving them plenty of attention, and they can reinforce – or sometimes even spark – children's willingness to achieve by providing a suitable learning environment. The latter is most successful when the challenges children are given coincide closely enough with their ability to learn for most of their efforts to be successful. To take reading as an example: the challenges posed by vocabulary, the complexity of sentence construction and the subject of the text should be slightly above the child's current level of development. If a child is challenged too much or too little, his motivation to learn decreases or disappears altogether. Adults react to being over-challenged in exactly the same way – at work, for example. Their willingness to learn is maintained when they have at least some degree of success.

Self-directed learning is the only learning that lasts

The opportunity to learn independently is just as important for children as facing challenges that coincide with their level of development. There are two strong arguments for this. For one thing, every child develops their own strategies for linking what they have just learned to the abilities and knowledge they already have. For another, learning experiences are only internalized in a lasting way when they can be woven into children's existing abilities, skills and knowledge. The psychologists Aljoscha Neubauer and Elsbeth Stern have this to say on the subject: 'from babyhood to old age, the same rule applies: successful learning takes place when the information being received is joined onto existing knowledge'.[12] And who is most familiar with the level of children's knowledge and abilities? Children themselves.

If children are allowed to learn independently, when reading, for example, they will work from their current reading ability and make an effort to read words and sentences that are slightly more challenging in terms of both form and content. But when children's learning is directed by someone else, they are much less able to join the experience that has been forced on them (a text, in this case) on to their existing

Lasting learning

Rote learning

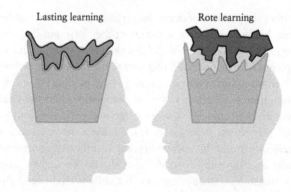

Figure 3.14: Lasting learning and rote learning. In learning that lasts, abilities, skills and knowledge (dark grey) are joined on to existing abilities, skills and knowledge (light grey) through development-specific experiences. Things that are learned by rote (dark grey) are only partly networked into existing abilities, skills and knowledge (light grey) – or not networked at all. Things learned by rote are therefore quickly lost again.

abilities, skills and knowledge. Things learned by rote and practised mechanically are quickly forgotten again – and, unfortunately, this is something that happens quite often in schools.

The things that children learn should endure. Academic success shouldn't be determined by children regurgitating the things they have learned by rote and practised; they should be judged on learning they have done independently. The curriculum and the school time-table shouldn't focus on exam grades, but on abilities, skills and knowledge that a child can use for years to come, and on into adulthood. How learning happens is just as essential as what is learned.

The principle that self-directed learning is the only learning that lasts also applies when childhood is over. When adults want to learn something, they need to be able to experience it independently in order to join what they have learned on to their individual abilities, skills and knowledge. Adults need to find out for themselves what interests them and what information and experiences they need in order to expand their competencies and knowledge. Rote learning of theories and practising formulaic processes, both of which are widespread in further education and training, don't add to competencies and knowledge in a lasting way.

A lack of emotional security impairs development

In the 1940s, the psychoanalyst and child development specialist René Spitz spent some time in children's homes and observed the damage done to children's development by emotional neglect and the lack of experiences that often comes with it.[13] The significance of emotional security and affection for development in the first years of life is also shown by a more recent study that tracked the progress of Romanian orphans.[14]

In 1966, the Romanian president, Nicolae Ceaușescu, set a target of raising the country's 'human capital' from 20 million to 30 million people by the start of the following millennium. The idea was that population growth would improve the country's economic output. Birth control and abortion were outlawed, and families with fewer than five children were punished with an 'abstinence tax'. The measures led to a huge leap in the birth rate. But many families were so poor that they couldn't afford to feed their children and were forced to hand them over to state orphanages. By the time Ceaușescu was deposed in 1989, 170,000 children were living in these institutions in the most miserable conditions. They weren't just psychologically neglected; everyday life in the orphanages offered them too few learning experiences. After 1989, the countries of Europe and a number of NGOs began a sustained intervention, training nursery nurses and improving conditions in these institutions. Many children were adopted by parents from the rest of Europe and the USA.

In Bucharest, scientists conducted a study of orphans, hoping to answer two questions: firstly, how damaging had the insufficient care in the orphanages been for these children's development and behaviour? And could being housed with a foster family mitigate or even compensate for developmental delays and behavioural disorders? In total, the study looked at 136 babies and young children from six large orphanages. At the start of the study the children were between six and twenty-two months old. They were divided into two groups. The children in Group A had to go on living in orphanages, while the children in Group B went to live with foster families. Conditions in the foster families were good enough; the foster parents received financial support and advice from social workers. Group C served as a control group, consisting of children who had lived with their biological families from birth.

All the children were assessed thoroughly several times. At the age of fifty-two months, the children in Group C had developed normally for their age (development quotient (DQ) 100). At fifty-two months old the children in Group B, who had been taken in by foster families, had a developmental age of forty-two months (DQ 80). At the same age the children in Group A, who had remained in the orphanages, had a developmental age of just thirty-six months (DQ 70). The researchers also discovered that orphans who had been taken into foster families before they were two had developed significantly better than those who had been fostered after their second birthday or even later. The improvement was particularly noticeable in children who were fostered at between six and twelve months old. The earlier the child was removed from the deprivation of the orphanage and started to receive sufficient encouragement, the better they developed.

There were striking differences between the three groups when it came to social and emotional development, too. The children in Groups A and B were almost three times more likely to suffer from emotional disorders and behavioural difficulties than the children who lived with their parents (53 versus 20 per cent). Almost half of the orphans in the foster families had formed a trusting relationship with their carers by the age of forty-two months. For the orphans who had to go on living in the orphanages, this figure was just 18 per cent (Group C: 65 per cent). This study confirms the observations that René Spitz had made sixty years earlier: a lack of emotional security, coupled with insufficient learning experiences, impairs children's cognitive, social and emotional development.

These developmental disorders aren't just seen in countries where poverty and social problems are rife. The following case study of a little boy, Tobias, comes from Switzerland – a wealthy and well-ordered country. It gave me a lot to think about. Tobias was a contented, active baby, whose development was above average. When he was two years

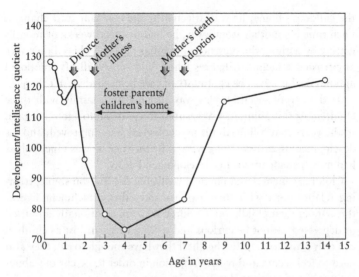

Figure 3.15: The progress of Tobias's development between the ages of one and fourteen, as measured by the development or intelligence quotient. A DQ/IQ of 100 corresponds to an average development: values above this represent accelerated development; values below represent delayed development. (source: Zurich Longitudinal Studies)

old his parents separated, and soon got divorced. Tobias's mother had custody of the boy. Six months after the divorce, she fell ill with breast cancer, and over the following four years she spent weeks or months at a time in hospital. During those periods when Tobias's mother was unable to care for him, he was placed with various foster families or in a children's home. She died when he was six.

His mother's illness and death plunged Tobias into a period of profound grief and listlessness. As we can see from his developmental curve, however, his development had already slowed down. At the age of four, a substantial developmental delay had set in (IQ 73; developmental age thirty-six months), even though all the people who had looked after Tobias had made a real effort with him. Why was the care they had provided still not good enough?

When his mother was in hospital, Tobias lived with a series of different foster families, often spending just a short time with each. In the children's home, he was affected by the constant staff changes. Each time his mother went away, he had to spend weeks or months getting by without the presence of familiar people. Tobias didn't have any trusted attachment figures who could satisfy all his emotional and physical needs. The continuity of care he needed wasn't there.

At the age of seven, Tobias was adopted by a family with three children. At this point, he was securely integrated into a family again. In the years that followed, his psychological state improved, and his development improved along with it. By the time he was nine, Tobias had mostly made up for his developmental delay.

Children's mental and physical wellbeing depends on stable, trusting relationships with their parents and other attachment figures throughout their childhood. Children who are emotionally unsettled or depressed cannot be curious and willing to learn. Just as children can only grow to their full height if they have enough to eat, they also need to feel secure and receive affection in order to be curious about the world and motivated to learn.

FUNDAMENTALS FOR
THE FIT PRINCIPLE

Follow your own developmental path

Education sows not seeds in you, but makes your seeds grow.
Khalil Gibran (1883–1931)

The human brain is a miracle of evolution, much like the lime tree with its spreading crown, which grows from a tiny seed. But in terms of complexity, its branches are nothing compared to the neural webs in our brains, woven by the countless experiences that have proved useful over the course of evolution.

Every child has their own developmental map and their own pace of development. Parents and teachers are proud of children who make good progress in maths and relieved by their eagerness to learn. They don't have to urge these children on. But when children show no interest in numbers, parents start to worry. Students who simply refuse

Figure 3.16: A lime tree.

to understand maths also put their teachers under pressure. Some parents and teachers feel they have to give additional encouragement to children whose achievements fall below their idea of the norm. They force skills and knowledge onto a child when he isn't ready for them; his development hasn't reached that level yet. The child knows his parents and teachers expect something from him that he 'should' be able to do, but he can't understand it yet. He becomes convinced that he can't learn or achieve anything, and as a result his curiosity and motivation to learn are impaired. Parents and teachers always know better than me, he thinks – I'm a failure. But developmental biology tells us that there is no progressing beyond our potential. Or to put it another way: if a child is malnourished, he won't grow as tall as he should. But overfeeding him won't make him taller; he'll just get fat. The same applies to cognitive development: if a child is neglected, he will develop less than he should. But overstretching him won't make him cleverer; he'll just get demotivated. Even the most talented educator with the most sophisticated teaching strategy won't get every student to achieve good grades. She can communicate the same academic material to all her students and set them the same challenges. But after a few years, the differences between those students' levels of attainment will be greater than ever before.

Figure 3.17: independent and enforced learning.

However, when children are allowed to learn independently, they are able to fulfil their learning potential to the greatest possible degree. As adults they will be aware of their talents, strengths and weaknesses. They will have a good sense of self-worth and self-efficacy, which are just as crucial to their wellbeing as realizing their potential is. But, much to children's disadvantage, some teachers, parents and education policy makers still refuse to accept this important insight: they have also seen that conscientious rote learning produces better grades. That may be true, but it doesn't make students more intelligent, and it decreases their motivation to learn.

As parents and teachers, how should we shape our attitude towards children? Just as a lime tree sapling doesn't grow faster if we pull on it, children won't develop better if we try to push them beyond their potential. One thing on which we can rely is that all children want to learn and achieve – but independently, in their own way and at their own pace. Our principal task is not to teach children skills and knowledge, but to enable them to have the experiences they need. We bring children up to be good members of society less through constant correction and strict rules of behaviour than through setting an example, demonstrating good interpersonal behaviour and moral values.

If a child is allowed to follow his own developmental path, he will become more and more himself. We all want to keep getting closer to our essential selves, at every stage of our lives.[15] From the Fit Principle point of view, self-realization is one of humanity's basic needs.

How experiences shape us

We grow through our parents, our siblings, our teachers, our experiences, our friends, our role models and our worthy opponents.

Wolfgang J. Reus (1959–2006)

Experiences allow children to develop their abilities and acquire skills, but they also provide information – about the culture in which they live, for instance. In Europe children eat with a knife and fork, in China they eat with chopsticks and in India with their hands. The way in which they absorb information is expressed particularly clearly in the development of language and social behaviour. Communicating with other people enables children to acquire not only the formal elements of language such as grammar and syntax, but also the semantic meaning of words and phrases. Communication is how children internalize the structural elements of human relationship behaviour, such as gesture and facial expression, and information, such as social rules and moral values. The ritual greeting that European children learn from their role models involves offering someone your hand and looking them in the eye. In Japan, by contrast, children learn to bow when greeting someone, and avoid eye contact and touching. Both rituals, although very different in their form, are an expression of respect.

One peculiarity of childhood experiences is that they both promote the development of existing structures and provide information. That means the development of structures and the acquisition of information go hand in hand. The information learned in childhood is very tightly woven into the maturing brain structures, which makes it far more enduring than anything we learn as adults. Memories like the smell of your grandmother's kitchen or the sound of bells from the village church can trigger a sense of deep familiarity even decades later. But memories connected to mental and physical violence can also reawaken intense fear and horror.

The things we experience during childhood shape us for both good and ill. Children's social behaviour is decisively influenced by the way their parents treat them. Deeply damaging experiences such as violence within the family can't simply be expunged from someone's (largely unconscious) memory later on; they are firmly tied to the structures of social behaviour. What we can do is try to replace troubling childhood experiences, at least in part, with new, positive experiences. Someone who gets to experience a trusting, non-violent relationship with a partner will suffer less and less from the painful memories of a violent childhood. Someone damaged by negative experiences at school can strengthen their sense of self-worth and self-efficacy through success at work.

The genuine need to develop talents

Wolfgang Amadeus Mozart was able to develop his musical genius to the full because he grew up in a family who gave him plenty of encouragement. His childhood was an ideal coalescence of extraordinary talent and proper support. The famous pianist Arthur Rubinstein's early childhood was a very different story. Arthur was born in 1888, the seventh child in a Jewish family of hand-weavers in Łódź, Poland. He grew up in a family where, in his own words, 'no one had even the slightest musical talent'. As a small child, Arthur spoke little, though he sang a great deal and was irresistibly drawn to musical notes and sounds. He loved working out well-known melodies on the piano keyboard. A respected piano teacher took the gifted boy on as a pupil, but had little success with him: Arthur fell asleep when he was supposed to be practising his finger exercises. Despite his less than encouraging environment, Arthur Rubinstein gave his first Mozart concert at the Philharmonic Hall in Łódź at just seven years old. His great musical talent won through even under less than advantageous circumstances.[16]

Wolfgang Amadeus Mozart and Arthur Rubinstein are exceptional cases. But talents will out in all children, not just the highly gifted ones – whatever form these talents take. Even children with learning difficulties are curious and want to develop their talents, though at their own pace and in their own way. Their efforts deserve our respect just as much as the achievements of gifted children.

But what about curiosity and motivation to learn in adults? In 1837, the 27-year-old Charles Darwin returned to England from his round-the-world voyage on the *Beagle*, after nearly five years. He was not in good health. He lived for another forty-six years, but never quite

Figure 3.18: 'In autumn 1766 the Mozart family visited Zurich on their journey across Europe. At the start of October, the ten-year-old Wolfgang Amadeus Mozart gave a demonstration of his extraordinary ability. On the concert flyer he was hailed not only as a prodigy on the piano, but as a "virtuoso in the art of composition". On the back of the flyer, Mozart wrote out a few bars of one of his own compositions.' (source: Zurich Central Library)

recovered from the mysterious illness he had picked up on his voyage. Darwin repeatedly found himself in such a weak state that he could hardly get out of bed for months at a time. But his physical difficulties didn't stop him working with a fierce discipline, analysing the samples of plants and animals he had brought back from his voyage.

He also carried out a large number of experiments using plants and animals native to Britain. His insatiable curiosity lent him immense strength. In 1859, at the age of fifty, he published his magnum opus, *On the Origin of Species by Means of Natural Selection*.

Scientists like Charles Darwin and great inventors like Thomas Edison spend their whole lives in search of knowledge. The drive to understand the world gives them the strength and the will to set themselves extraordinary challenges. Artists like Arthur Rubinstein and Pablo Picasso never give up their art, either. They are simply incapable of stopping. Describing this driven quality, the composer Arnold Schoenberg said: 'I believe art comes not from ability, but from need.' Scientists and artists alike find the meaning of life in the exercise of their talent. The temple built by the Swiss graphic artist and sculptor Karl Bickel (below) represents a life's work literally set in stone. Like all characteristics, the drive to create differs greatly from one person

Figure 3.19: 'Paxmal': a monument to creative effort. In 1924 Karl Bickel, a Swiss graphic artist and sculptor, began constructing a 'temple to peace' in a rocky field on the Walenstadtberg. He was twenty-seven at the time, and spent twenty-five years working on his creation without any outside help, finally completing the temple in 1949.

to the next. Some people are exceedingly creative and productive in their work and hobbies right into old age, while others lose the desire to create by the age of twenty.

Different ways to get smarter

Once adolescence is over, all the body's organs are fully developed, including the brain. Its structures are now mature and therefore largely fixed. Some really remarkable things are achieved in puberty and in the ten years that follow. The great mathematical discoveries have almost all been made by people between the ages of fifteen and twenty-five, and often before the age of twenty. They are not the fruits of a decades-long career, but the products of an innate intellectual capability, which the psychologist Raymond Cattell has described as 'fluid intelligence'.[17] This intelligence encompasses logical and mathematical thinking, problem-solving strategies and figural and spatial thinking.

Blaise Pascal (1623–62) invented the first mechanical calculator at the age of nineteen. He developed it to support his father in his work as a royal inspector and chief tax collector. It is said that Carl Friedrich Gauss (1777–1855) started helping his father produce pay slips at the age of three. His father was a gardener, slaughterman and treasurer for a small insurance firm. At fourteen, Gauss produced his first thoughts on non-Euclidian geometry. At eighteen he developed the 'method of least squares', which gave rise to the Gaussian bell curve of normal distribution. A year later he succeeded in proving that it was possible to construct a regular seventeen-sided shape, a problem which had been eluding mathematicians since ancient times.

Albert Einstein published his special theory of relativity and his famous formula, $E = mc^2$, at the age of twenty-six. He then expanded on this work to produce the general theory of relativity. In the decades that followed, he continued to work avidly in the fields of mathematics and theoretical physics, but he didn't make any more great discoveries. The impressive advances in information technology of the past decades have not been achieved by experienced minds in grey-haired heads, but by young programmers who are often still teenagers.

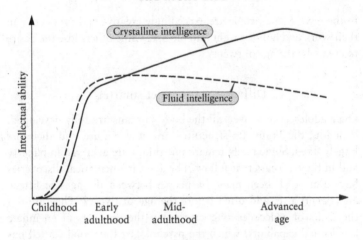

Figure 3.20: Fluid and crystalline intelligence. Fluid intelligence reaches a peak in early adulthood and then declines slowly and steadily. Crystalline intelligence continues to grow until an advanced age and then also declines.

Another area of development that peaks during adolescence is motor agility. The Romanian Nadia Elena Comăneci is famed for having been one of the best gymnasts of all time. At the summer Olympics in Montreal in 1976, when she was just fourteen, she became the first gymnast ever to score a perfect 10 on asymmetric bars. Until that point, people had thought the highest mark was impossible to achieve. Comăneci also won gold on beam and in the individual all-around competition, bronze on floor, and silver in the team all-around. At eighteen she won gold on beam and floor again at the 1980 Olympics in Moscow, plus silver in the individual and team all-arounds. Top performances in other sports that demand a high degree of motor agility, such as figure skating or tennis, are also most often given by athletes under the age of thirty.

People may find it difficult to accept that once puberty is over the brain has largely finished maturing. From then on, we can no longer learn the same way we did in childhood. Our abilities – the way we walk, for example, or the way we use words or facial expressions – are mostly fixed. As adults, we just have to live with the way our brains have developed. But there is still hope for us, in the form of crystalline intelligence.[18] This term refers to an ability to learn based on gaining experiences and knowledge. A doctor, for instance, can gauge the significance of symptoms better and more accurately after years of experience than a younger, less experienced colleague can. Brain maturity, and with it the creation of neural networks, is complete by the time we reach adulthood. But synaptic connections can be made and broken throughout our lives, depending on the experiences we have and the knowledge we gain. But, as in childhood, adults can't make use of just any old experience: we need those experiences that we can join on to our current level of skill and knowledge. This is how we improve our IT skills, for example. We can also improve our social competencies through spending time with members of our family and dealing with work colleagues. It is still possible for people to put in top performances – though of a different kind to those produced by fluid intelligence – at an advanced age. Poets and musicians, painters and sculptors, are inspired to create new and sometimes great works by the many and varied experiences they have over their lifetimes. Johann Wolfgang von Goethe completed his later work, *Faust I*, at fifty-nine and *Faust II* at eighty-three, just a few months before his death. Even when our eyesight and hearing gradually start to fail, and our limited motor function makes us unsteady on our feet, experiences can still make us smarter – or perhaps even wise.

4
Basic Needs that Shape Our Lives

Each human has a unique profile of needs.

> *We hold these truths to be self-evident, that all men are created*
> *equal, that they are endowed by their Creator with certain*
> *unalienable Rights, that among these are Life, Liberty and*
> *the pursuit of Happiness.*
>> United States Declaration of Independence (1776)

For the founding fathers of the United States of America, the 'pursuit of happiness', as written into the preamble to the Declaration of Independence, was a basic right, just like the rights to life and liberty. People want to be happy. But what is happiness? Philosophers and religious leaders have been wracking their brains over the question for at least 4,000 years. In the last 100 years psychologists, doctors and, more recently, neuroscientists have also started to look into the subject of human happiness. And in the last few years there have been several high-profile studies in which economists and social scientists have explored the link between subjective satisfaction, income and wealth.

In order to measure satisfaction, some of these scientists asked test subjects for a subjective assessment of their state of mind, which went beyond discrete areas such as health or working conditions and covered as many aspects of life as possible. They wanted to measure feelings and moods such as confidence or fear, along with more objective factors including standard of living and professional status. As the graph shows, the study proved one significant connection: subjective life satisfaction grows with increasing income, very steeply at first, in

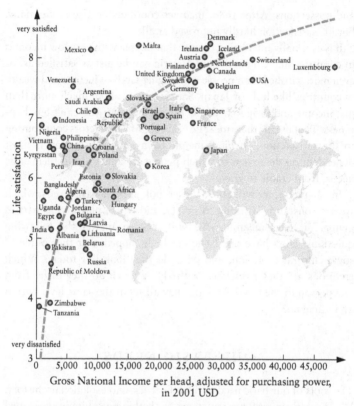

Figure 4.1: Relationship between life satisfaction and gross domestic product (GDP) in sixty-three countries. Scale of life satisfaction: 1 = very dissatisfied; 10 = very satisfied. (source: World Values Survey 2001)

developing countries such as Tanzania for example, and then much less steeply in highly developed countries like the USA. As incomes grow, life satisfaction increases at a much slower rate. In Western countries, income and wealth only contribute about 10 per cent to life satisfaction. The same trend was seen in Germany in the years after the Second World War. During the period of the 'German economic miracle', life satisfaction increased rapidly as incomes went up. With rising incomes came an increase in material goods and improved living conditions, including healthcare provision and security for people and possessions. After 1980, incomes continued to rise in Germany, but life satisfaction hardly improved at all.

It is especially interesting to note that people with very low incomes in countries like Colombia and Nigeria can be just as satisfied as, or even more satisfied than people with higher levels of income and wealth in countries like Italy or Japan. Satisfaction is about much more than just income and property. Or, as the saying goes: you can't buy happiness. That means there must be other significant factors contributing to life satisfaction. In Fit Principle terms, life satisfaction is most dependent on the extent to which people are able to fulfil their basic needs, from cradle to grave.

In Chapter 4, we will take a closer look at these basic needs, which we are constantly working to satisfy, and which play a vital part in giving our lives meaning. We will seek to answer the following questions: How have our basic needs developed over the course of socio-cultural evolution, and what do they look like today? Which emotions are they associated with? How much do they differ from one person to the next? And are they all set at the same level within an individual?

OUR BASIC NEEDS

The roots of our basic needs stretch far back into evolutionary history. We share their preliminary stages with the more highly developed animals. All animals want to satisfy their hunger and thirst. Like human children, young chimpanzees have a great need for emotional security and affection. Foxes and wolves fight for their social status within the pack. Without exception, young animals want to realize their abilities – motor functions, for instance. Some species perform highly specialized tasks: honey bees build artful combs out of wax, and leaf-cutter ants cultivate vast fungus gardens. Some animals, such as squirrels or dormice, lay down stores to provide for themselves over winter.

Animals aren't conscious of their needs, and the process of satisfying them is guided by emotions such as sexual desire. We humans believe we are aware of our basic needs: I'm tired, so I'll go to bed.

But our basic needs are very old in evolutionary terms, and as such they are deeply anchored in our unconscious. This means we are often only partly conscious of them, or not conscious of them at all – as with our need for social recognition and social status, for example. If you ask a mayor why he applied for the position, he will in all likelihood say it was because he wanted to make a (completely selfless) contribution to the community.

Another thing that separates us from other animals is that we have made huge strides in how we satisfy these needs. The ways in which we do this – the production of food, for example – are almost as important to us as the satisfaction of the need itself, and sometimes even more so. Over the last 10,000 years a cultural transformation of needs has taken place. Emotional security is no longer just a strong need; the theme of 'love' rules the worlds of literature, music and film. Creativity, the drive for self-development, has been fertilizing culture for thousands of years, and the innovations it produces power science, technology and the economy. In order to protect people and property as well as we can, we have developed an elaborate set of constitutions and laws, as well as state institutions like the police force and the army. Ultimately, our attempts to satisfy elemental needs have produced an enormous wealth of knowledge and ideas.

Figure 4.2: The six basic human needs.

The Fit Principle works on the basis that human lives are largely determined by six basic needs. We will flesh out these needs in the chapters that follow.

PHYSICAL INTEGRITY

For humans to feel good physically, we need to get enough to eat and recuperate through sleep; we don't want to shiver with cold or sweat excessively; we need to be able to express our sexuality, and to be healthy and physically capable. Only when our physical needs are met can we start to satisfy the other basic needs.

All humans, from newborns to the very elderly, want to do more than just satisfy their hunger and thirst. They also devote a lot of time and effort to how they do this. They fry, boil and spice their food; they ferment fruit and vegetables to make alcoholic drinks. On special occasions, meals become a celebratory social event with elaborate table settings, speeches and singing. Animals protect themselves from the cold and wet with fur and feathers. The naked human does more than just protect himself against bad weather with skins, furs and various textiles. He also manages to adapt to a wide variety of climates, and makes clothes according to his needs and an ever-changing array of aesthetic criteria. Animals groom the fur or feathers that clothe them, to get rid of the dirt and parasites that might make them ill. Humans have extended this utilitarian exercise in maintaining hygiene to incorporate baths and massages, the use of soaps and essential oils, transforming it into a whole spa and wellness industry. It is a long time since the human sex drive served nothing but reproduction. Eroticism has played an essential role in human life for thousands of years, going far beyond its original biological significance and becoming an integral part of every culture. Eroticism finds expression in art and is marketed in countless different ways by the fashion industry. The introduction of the contraceptive pill fifty years ago decoupled sex from reproduction and led to increased economic exploitation of eroticism and sexuality.

Health is an essential component of physical integrity. Even tens of thousands of years ago, our ancestors did their best to ward off illness. Shamans came to the aid of the sick, carrying out healing rituals and calling on the gods of nature and supernatural forces for help. Female healers used herbs and minerals, massages and baths to alleviate suffering. Modern medicine as we know it is a mere 150 years old, though in this time it has been extremely successful in the diagnosis and treatment of illnesses. Vaccination programmes have eradicated severe infectious diseases such as smallpox and polio. The huge overall improvement in human health is also reflected in our average lifespan, which is now double what it was 150 years ago. But our awareness of health and expectations of medicine have increased along with it. More and more people are now trying to adopt a healthy lifestyle, following diets and getting fit. Health and physical integrity are no longer just the luck of the draw; they are at once an aspiration and an individual responsibility.

The diversity of linguistic expressions for physical needs and the emotions that come with them tells us how significant they are. We say, for example, that we could eat a horse or that our stomachs are rumbling. When our mouths are dry, we are 'dying for a drink'. Lust and climax send us into ecstasies. After a hard day at work we feel drained, and after a good sleep we are full of beans. On good days we feel hale and hearty, on bad days miserable and flat. Some people fire on all cylinders, while others are stressed or even suffer from burnout.

Terms and emotions we use to express our physical condition
Food and drink
Starving, insatiable, stuffed; cravings, rumbling stomach, famished
Gasping or dying for a drink, drinking greedily; parched
Sated, full to bursting
Sleep
Exhausted, sleepy, dog-tired, wiped out
Well-rested, refreshed, on form, full of beans
Sexual desire
Passionate, stormy, lovesick, lecherous; surrender, climax, ecstasies
Sated, satisfied, listless

Health
Hale and hearty, blooming, unscathed, robust, unspent
Suffering, care-dependent, mortally ill, injured, disabled, fragile
Fitness
Firing on all cylinders, fit as a fiddle, on top form
Flat, feeble, stressed, exhausted, burnt out

In the first years of life, children are completely dependent on their parents and carers for the satisfaction of their physical needs. But the older they get, the more of a say they want to have in how their needs are satisfied – in what they eat and drink, for example. Even so, they only become fully independent as teenagers. At this point, most young people start to decide for themselves what they want to eat, drink and wear. They have their first sexual experiences, mostly without their parents' knowledge. Once they are adults, they can choose from an incredible range of consumer products with which to satisfy their basic physical needs. But they also bear an individual responsibility – for how they feed themselves, for example, to avoid becoming overweight and developing diabetes.

Humans don't cope well with being able to satisfy their needs to excess, by eating as much as they can, for example, or having sex as often as they want. They feel much better when they are able to satisfy their individual needs to the level they require. There are people who place great importance on food and drink, and others who have very little interest in them. Sexuality plays a central role in some people's lives, and a more marginal role in the lives of others. Some people go to the gym several times a week, while others are baffled by the idea. However your need for physical integrity may be weighted, satisfying it is an important precondition for being able to satisfy other basic needs, such as the need to achieve.

EMOTIONAL SECURITY
AND AFFECTION

We share our elementary physical needs with other animals, and we have common psychological needs, too. One ancient need is the desire for emotional security and affection. Over the course of evolution a lot of animals – in particular birds, mammals and humans – have developed 'attachment behaviour', which comes with a great desire for intimacy and affection (see Chapter 5). In parallel to this, parents have developed nurturing behaviour. Young animals and human children wouldn't survive without forming attachments and being cared for by their parents. Children develop a strong emotional dependence on their parents and other carers, which ensures that their physical needs for food, care and protection are reliably met. Neglect can have a huge impact on their growth and development (see Chapter 3). And human children don't rely on emotional security and affection for a mere few weeks or months like young animals do, but for at least fifteen years.

The emotions that come with security and affection express our longing for unconditional acceptance. They are some of our strongest feelings. For the majority of people, love is the most important emotion of all. We link a wide variety of ideas to it: filial, maternal and paternal love, romantic love, selfless love for our fellows, and even the love of God. Negative emotions such as jealousy and hate are also

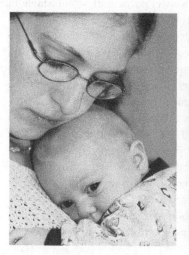

Figure 4.3: Feeling emotionally secure.

connected to the longing for emotional security. These are the feelings that feature most often in literature, theatre, film and music.

How we express our need for emotional security
Feeling secure and unconditionally accepted:
Maternal/paternal love, fidelity, warm-heartedness, devotion, intimacy, attachment, affection, tenderness, emotional security
Feeling rejected, neglected, abandoned:
Suspicion, jealousy, hatred, resentment, mistrust, rejection

The need for security and affection develops in the first weeks of life and is maintained throughout childhood. In adolescence, these ties are loosened enough for a young adult to be able to leave the family and enter into a romantic relationship with someone else. As adults, we still rely on security and affection, though less than we did as children. A sense of belonging is important to our mental well-being, as well as intimacy and affection from people we trust. We cannot be alone long term, and we feel particularly comfortable in familiar surroundings that give us a sense of home. No matter what age they are, children aren't all dependent on emotional security to the same degree. There are two-year-olds who have no difficulty saying goodbye to their parents at nursery and feel they are in good hands with the nursery staff. Other children still find it difficult to be apart from their mothers by the time they start school. Even ten-year-olds can be reluctant to spend a week away on a school trip. Adults also have different levels of need for intimacy and affection. The desire for the emotional security provided by marriage, family and community varies from one person to another. Some people experience being abandoned – in a divorce, for example – as a tragedy, while others will be much less affected (see Chapter 5, 'Social behaviour'). When it comes to our ability to satisfy other basic needs, getting enough emotional security and affection is just as important as being well-fed and healthy.

SOCIAL RECOGNITION AND
SOCIAL STATUS

In the long term, many animals can only survive in a community. This is true of both simple creatures like ants and more highly developed animals like chimpanzees. Communities are particularly crucial to humans (see Chapter 1). Only a tightly woven network of interpersonal connections, mutual dependence and interests can give humans a sense of emotional, social and existential security. Living in communities, our ancestors also developed a need for social recognition and a secure social status. Along with the desire for emotional security and affection, this need made humans into highly social animals.

Social recognition and rejection, high and low social status, come with a great many emotions. Some of these are listed below.

Terms and emotions that express recognition and social status
Feeling valued, respected, honoured, striving for prestige and status:
Recognition, respect, deference, reputation, authority, rank, esteem
Feeling undervalued, rejected, bullied:
Exclusion, lack of respect, humiliation, contempt

The respect we enjoy from our family, friends and colleagues, and the status we occupy among them, is vital to our sense of wellbeing. The enormous significance of how we are perceived socially is laid out for us day after day in emotive media reports. Criminals and failures are portrayed with loathing and rejection; prominent figures in politics and showbusiness are showered with almost hysterical forms of attention and admiration, and extraordinary people like Mahatma Gandhi or Nelson Mandela are regarded with deep veneration and esteem. The significance of respect and status led to the formation of hierarchical structures within communities very early in human history. Tribal chiefs may well have existed as early as 200,000 years ago. People were governed by pharaohs, kings and noblemen for thousands of years; serf-dom and slavery were widespread. With the Enlightenment, equality became a basic right in Western civilization, but it has never been wholly implemented, as we can see from the gender pay gap among other things. And many countries still have rigid, discriminatory social structures – India, for example, where people still live in a caste system from which the 'untouchables' have a hard time escaping.

Even toddlers have a desire for social recognition within their families and nursery groups, wanting to occupy a position that suits them. Do my siblings like me? Do the other children at nursery accept me, and do they want to play with me? When they start school, the need for recognition and a secure position in their peer group continues to grow, becoming all-important during puberty. For some young people, their friendship group is more important to them than their own family. As adults, just as we have seen with the need for emotional security, social recognition and social status are more important to some people than to others. One person might seek recognition above all within their family and social group and place less importance on their status in the workplace, while for someone else the reverse might be true. And for some people social recognition is equally important in their family, social group and workplace. People don't have to reach the pinnacle of a social hierarchy, becoming a professor in a university or the CEO of a large company, before they feel happy. Most people are content with taking up a social position that suits them and receiving the recognition they can expect for their level of ability. But all people, no matter how modest their social position, want to be respected. All forms of exclusion damage their wellbeing. Unfortunately, more and more people in our society quite rightly feel as if they receive too little social respect (from their relatives, for example), and are not able to take up a position that suits them (for instance, at work).

SELF-DEVELOPMENT

All animals, without exception, feel a need to develop their innate abilities to the highest possible level, and this need is particularly pronounced in humans. Children have a strong desire to develop their abilities and acquire skills and knowledge. But, unlike other animals, humans retain this desire into adulthood (see Chapter 3).

From early childhood, developing our talents is associated with positive feelings such as curiosity, ambition and pride, but also with negative feelings including frustration and discouragement. The strength of these emotions gives us an idea of just how important self-development is for humans.

Terms and emotions associated with the development of talents
Realizing yourself, learning, mastering something, reaching a new stage of development:
Curiosity, pride, thirst for knowledge, urge to explore, creativity
Failing, falling behind, becoming disillusioned or disappointed:
Frustration, inability, failure

During the first years of life, children become increasingly aware of the developmental progress they are making, and their emotions reflect this. They are pleased and deeply satisfied when they reach milestones in their development such as being able to walk unaided. But they also know when they haven't succeeded in doing something. An eighteen-month-old child is very proud when he manages to build a tower of blocks. But if his tower falls down, he may throw a tantrum.

When small children have developed sufficiently, they want to use the toilet and put on their clothes and shoes themselves. Once they've started school, all children want to learn to read and do maths, albeit at different ages (see Chapter 3). During adolescence, the brain reaches maturity, and at this point the motivation to learn is largely extinguished; mature cats only chase balls of wool to satisfy their hunting instinct. In humans, however, a certain amount of curiosity and willingness to learn is retained even after adolescence. Some adults go on wanting to expand their knowledge and skills up to an advanced age. They broaden their knowledge of a foreign language or continue to develop their talent in painting or pottery. Even if the progress they make is limited, it still gives them a strong sense of satisfaction.

Humans of every age want to develop their individual talents in their own way and at their own pace (see Chapter 3). A boy may be able to ski at the age of two, even though he can't really form words yet. His sister may be able to speak in whole sentences at two, but won't venture on to the piste until she's five. There are adults who spend their whole lives struggling to write a novel and hoping for a breakthrough. Others don't feel any need to create something new and unique.

Children have high, often unrealistic expectations of themselves. Six-year-old Theresa, a child whose motor skills are no better than average, wants to be a prima ballerina, and Konrad, an eight-year-old who is short for his age, wants to be a goalkeeper. Adults, too, are always hoping for things that never happen. And yet most of us aren't unhappy in the long term. An amateur painter can be thoroughly satisfied with his work even if it's a long way from being as good as that of his role model, Pablo Picasso. Ultimately, our wellbeing and sense of self-worth don't depend on how successful we are, but on whether we fully develop our individual talents. This is the real stimulus for self-development, much more than social recognition, even though the latter is something we enjoy receiving.

STRIVING FOR ACHIEVEMENT

When honey bees build artful combs out of wax or leaf-cutter ants cultivate their fungus gardens, do they experience the feeling of a job well done and waggle their feelers with joy? We don't know. But what we do know for sure is that, for humans, achieving things is a basic need. Humans don't work just to make a living or to attain professional success, social recognition and social status. Humans also want to achieve things, because achievement contributes to our sense of self-worth and self-efficacy. In areas like education, work or free time, people want to accomplish something they can be proud of. But everyone strives for success in their own way. Parents put all their energy into bringing up their children; a florist tries to create the most beautiful bouquets; and a company boss and her colleagues want to make their business a financial success.

What is the difference between achievement and self-development? Self-development means working on abilities such as understanding numbers and acquiring skills such as playing tennis. Achievement means using the abilities, skills and knowledge you have already gained to reach a particular goal. A railway engineer, for example, wants to use his technological know-how to make the trains run as smoothly as possible. Depending on what you are doing, achievement and self-development can overlap, as when the engineer learns new IT processes in the course of his work.

Once again, our emotions tell us how important achievement is to us. If we accomplish something impressive, we are delighted, but if we fail at an important task we are devastated. If we are under-challenged, we feel bored and frustrated. If we are permanently overstretched, we feel discouraged or even exhausted.

What we experience when we try to achieve things
Achieve, accomplish, master, complete
Satisfaction and pride, taking pleasure in work, conscientiousness, profit, success; feat, vigour, masterpiece, victory
Fail, give up, spread yourself too thin, founder
Failure, discouragement, disappointment, mishap, defeat, burnout

---

Even as children, we don't just want to develop our talents: we want to achieve, and to do so as independently as possible. When a child uses a spoon to eat all her porridge by herself, she feels a deep sense of satisfaction and self-affirmation: I did it. The child is visibly proud of her achievement. She doesn't need recognition from her parents and admiration from her playmates in order to feel this pride. But if she gets both of these things as well, it feels even better. She will be particularly pleased if other children emulate her. Schoolchildren are deeply satisfied when they achieve not just what parents and teachers expect of them, but what they expect of themselves. But if they're overstretched because they haven't reached the stage of development required for the task at hand, they feel like failures, and their desire to achieve decreases.

For most adults, achievement is more important than self-development. We don't just work for financial rewards. Accomplishing things for ourselves and the community strengthens our sense of self-worth and our self-efficacy. And that means unemployment and failure to achieve aren't just experienced as an existential threat and social debasement: they also have an impact on someone's self-worth and self-efficacy. Most people wouldn't actually be pleased if they no longer had to work and could spend their whole day sitting around at home. Those facing long-term unemployment are deeply unhappy, and it can affect their mental and physical health.

People who achieve extraordinary things receive a lot of media attention – take Reinhold Messner, for example, the first mountaineer to climb all fourteen peaks over 8,000 metres, including Mount Everest, K2 and Annapurna. But many people achieve great things without the public hearing about it. A mountaineering friend of mine has climbed to all 350 mountain huts in the Swiss Alps, but this is something only his relatives and close friends know.

People achieve all kinds of things, both large and small – even extreme things like cycling all the way round the world – quite simply because they pose a challenge and make them feel proud. Year after year, around 50,000 runners take part in the New York Marathon. Only one man and one woman can win – but most participants are satisfied if they manage to achieve what they can realistically expect of themselves. You don't need to accomplish extraordinary feats to lead a fulfilled life. Most people just want to achieve the things their abilities and skills will allow.

EXISTENTIAL SECURITY

Over the course of evolution, many animals have learned to safeguard their existence. They protect themselves from bad weather and the threats posed by other animals, and they take precautions. The mole retreats into its underground passages, and the squirrel lays down stores of food. For hundreds of thousands of years, our early ancestors lived in caves, primitive tents and rudimentary huts. When they became settlers, they started building houses of wood and stone. With the rise of arable farming and cattle rearing, they erected barns to hold surplus grain and stables for their animals. Single farms became villages and towns and cities, which expanded steadily over millennia. Today, 38 million people live in the megacity of Tokyo – a population greater than that of some countries. In 2007, for the first time more people lived in cities than in rural areas, and by the year 2050 this proportion is predicted to rise to over two-thirds. Caves and huts have become luxury apartments.

Since humans became settlers, their labour has reaped larger and larger rewards, which now go far beyond mere subsistence. The surplus is turned into ever greater material wealth. Worldwide consumption has reached an almost worrying level in recent decades.

Our ancestors learned early on to secure their property against natural disasters, theft and enemy attacks. They fortified their settlements with fences, city walls and moats. Devices that had originally been used for processing food and hunting were turned into weapons. Knives became swords and hunting spears became lances. Soon, these weapons were being used not only to protect people and property, but in conquests, raids and wars. The progress in weapons development, with the supposed intention of ensuring our existential security, has today been taken to absurd extremes, as epitomized by the nuclear bomb. The original need for existential security has been perverted into a global threat not only to humans, but to all life on earth.

Terms and emotions we use to express our need for existential security

Earning a living:

Breadwinner, income, right to work, social security, the welfare state, pensions

Being unemployed or made redundant, signing on, living rough; unemployment, being on the breadline, poverty

Shelter:

Feeling secure, protected, a sense of belonging; refuge, accommodation

Feeling excluded, displaced, ostracized, alienated, uprooted

Possessions:

Feeling well provided for, prosperous, rich, financially powerful, rolling in it; greed, luxury, avarice, miserliness

Being without means, miserable, in need; existential fear, envy,
resentment
Protection:
Feeling people and property are protected; legal certainty
Feeling threatened, endangered, unprotected, defenceless, without
rights or power; lawlessness, violence, persecution, war

People who earn a living for themselves and their families have long
been described as 'breadwinners', which shows what a great significance
the provision of food once had. Today we spend only a fraction of our
income on food, and many times more on accommodation and travel.

Rich people are accused of greed, but they are also envied. We are
sympathetic towards people living in poverty and facing homeless-
ness, and we want to help them. People who can't make a living for
themselves and have no property feel not only unprotected but as if
they have no rights or power. When we don't have enough existential
security, or have lost it entirely, we feel an overwhelming fear. Day
after day, the media tells us about the millions of people whose physical wellbeing is badly damaged by poverty and hunger, war and
displacement; people who are suffering from deep emotional insecurity and psychological trauma.

The need for existential security is set at very different levels in different people, and is always also an expression of their personal and
social circumstances. For some people, existential security princi-
pally means making a living and being protected by the rule of law;
for others it means wealth and power. Some people are very concerned about being looked after in their old age. Others give little
thought to it, hoping that fate will smile on them and the state will
find itself in a strong financial position when they reach that point.

Parents ensure their children's existential security. It is only in adolescence that children gradually learn to become independent and take
responsibility for making ends meet. Up until around fifty years ago,
most people moved out of their parents' house and started to take
responsibility for themselves at the age of about twenty. Today, this
process takes much longer, as we spend more time in education.
Increasing numbers of young adults are frightened by the enormous

outlay required for them to become existentially independent and stand on their own two feet. Often they only find themselves in a position to take care of themselves and start a family at the age of thirty or forty. The difficulty is compounded by the fact that even in developed countries like Italy or Spain, more and more young, well-educated adults are unable to find work because of the poor state of the economy. Guaranteeing existential security for everyone has become a huge challenge for society (see Chapter 10).

A UNIQUE PROFILE OF BASIC NEEDS

Basic needs are set at different levels not only from one person to another, but within the same person. Some people have a great appetite for eating and drinking, but little interest in sex. Young people no longer need as much affection from their parents, but social recognition from their peer group is very important. Some artists and scientists put a lot of time and energy into making a breakthrough in art or research, but place little value on public recognition. Other people just want to earn as much as they can, and don't care what sort of work they have to do to achieve this.

The different levels to which needs can be developed in individuals is illustrated by the following two profiles of basic needs, belonging to Jakob and Hannes. Jakob trades in real estate. He has his own company, and is married with three children. Hannes is an administrator. For the past ten years he has been working as the personnel manager in a large supermarket. His main interest is sport, and he lives with a partner who shares his interest.

Hannes' and Jakob's profiles reveal a number of differences:

Physical integrity: Jakob hardly gives any thought to his health. He is short and overweight and takes little exercise. Hannes, on the other hand, is in excellent shape and is a very able sportsman; physical fitness and health are extremely important to him.

Emotional security: in order to feel happy, Jakob needs a lot of emotional security and affection from his family. Hannes is much less reliant on affection in his relationship.

Social recognition and social status: Jakob sees himself as the head of the household. Recognition and social status are very important to him within his family, but less so in public. Hannes mainly seeks recognition in the world of sport. He is always pleased when his successes bring attention from the media.

Self-development: Jakob's need for self-development is low. He doesn't have any hobbies. Hannes, by contrast, puts a great deal of effort into improving the motor skills he uses in long-distance running and orienteering.

Achievement: Jakob is a workaholic who often works evenings and weekends. He also expects his children to achieve very good grades at school. Hannes does his job well, but he isn't looking to climb the career ladder. Success in sport is much more important to him.

Existential security: providing the greatest possible financial security for himself and his family is Jakob's real *raison d'être*, and it is what he puts all his time and energy into. He has amassed a considerable fortune. A secure income is important to Hannes, but he has no material ambitions beyond this.

The levels of our basic needs shape us as individuals and have a huge influence on our lives (see Chapters 8 and 9). Everyone has their own unique profile, which means we can't lead just any life. Hannes would have difficulty taking over Jakob's role as a real estate salesman. And Jakob would have even less chance of becoming a successful long-distance runner, even if his parents had put him on the right training programme as a child.

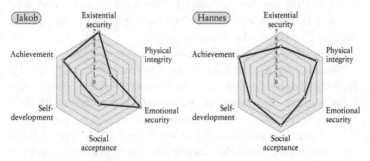

Figure 4.4: Individual profiles of basic needs. Needs are rated on the following scale: 1 = very low; 4 = average; 7 = very high.

Our genotype creates the organic basis of our needs and fixes the levels at which we experience them. Whether a child can satisfy her individual basic needs sufficiently depends on her environment. If she doesn't have enough emotional security – due to the death of her mother, for instance – her emotional state can be severely affected. But if that child is cared for well by her father and other attachment figures, such as grandparents, her mental and physical wellbeing won't be damaged in the long term. Depending on the levels of children's basic needs and their experience of school, they can either internalize the idea of achievement and social recognition as worthy aims or accord them little significance. The mindset formed at this stage will have an impact later on, in the world of work, where people can either take a very performance-focused attitude or just do what is required of them. If a child's parents strive for material gains with all their might then, depending on her profile of basic needs, a child may take on their attitude, be indifferent to it or even reject it. The child plays an active part. Her individual profile of needs triggers selective behaviour. Even if she grows up in a family with little interest in music, a musically gifted child will still want to develop her talent, much to her parents' amazement. She may attend violin lessons voluntarily and with great enthusiasm, and play in a youth orchestra.

All humans are anxious to satisfy their basic needs. If they succeed, they feel happy and grounded. If they don't – be it through their own fault or because the environment prevents them – they are dissatisfied and feel they aren't in control of their lives. Every day, they try again to satisfy their basic needs as well as possible. There is no set of generally applicable rules for doing this, because everyone has their own profile. The expectations and demands placed on us by parents, partners or society can make finding our way easier, but they can sometimes make it more difficult. An exaggerated pressure to achieve can make an employee feel insecure and start to question himself. What is more important to me: achieving things, or gaining social recognition and status? Or am I more attracted to a high income and wealth, and the admiration they draw from my neighbours? The process of finding out who we really are is a life-long effort to satisfy our basic needs.

FUNDAMENTALS FOR
THE FIT PRINCIPLE

How we satisfy our basic needs

Has not nature made us feel our needs as a means to our preservation!
Jean-Jacques Rousseau (1712–78)[1]

Our physical and mental wellbeing, in fact our very survival, depends on whether we can satisfy our basic needs adequately. The extent to which we achieve this and the way in which we achieve it constitute a large part of the meaning we give to life. When we look back in our old age, we want to be able to say: I have managed to satisfy my physical needs at all times and have been mostly healthy. I have experienced stable and trusting relationships, and felt emotionally secure and accepted. I've been able to develop my abilities, learn skills and achieve things that I found satisfying. And I have never had to face existential hardship. When we take stock of our lives, we also realize that satisfying our basic needs is a life-long challenge that we face every day. Without exaggerating, we can say that this effort occupies us twenty-four hours a day – even when we're asleep, we are meeting an elementary need. It is worth each of us making a thorough examination of how our own basic needs are constituted.

We can only lead our own life, not anyone else's. We have to accept our basic needs for what they are. Over the course of a lifetime, they pose a whole range of different challenges. Satisfying our physical needs is important at any age, but we do it in a variety of ways. Children want to grow and have a great need for food. Sexuality still has no meaning to them, though it plays an important role when they become adults. In order to develop, children need to feel emotionally secure and receive enough attention. As adults, this need is less pronounced, but the desire for a sense of belonging, social recognition and status is much greater. All children want the chance to develop their abilities, and some adults still want to satisfy this need through creative work. Children want to achieve things, though this need is much stronger in adults. Children don't worry about existential security, but once they become adults it will remain a concern for the rest of their lives.

Why do basic needs differ so greatly from one person to the next? Our genotype and the experiences we have in childhood both play a role here, but the most important factor is our *competencies* – things like language or mathematical thinking, which are discussed in detail in Chapter 5. The need to keep our bodies in shape, for example, is essentially dependent on gross motor skills. The better these are, the more of a need we feel to jog regularly or do some other physical activity. How much emotional security children want largely depends on their attachment behaviour. How hard a person strives for social recognition and social status is determined both by their genotype and by social experiences in childhood. Children whose main role models are their parents often have a difficult time if they have not inherited their parents' personality and strengths. Our need to develop our potential depends on the level of our competencies. Schoolchildren whose understanding of numbers is limited have a correspondingly low motivation to learn maths. Artists and scientists with greater than average creative or analytical abilities want to exploit those abilities all their lives. If we want a better understanding of basic human needs, we also have to take our competencies into consideration.

How do we fulfil our basic needs? The circumstances in which we live create the conditions for satisfying them, but the job of satisfying them is something we have to do ourselves. With physical needs, we are above all driven by feelings such as hunger. To satisfy their need for emotional security, children turn to their parents, who give them love and attention. Children and adults use their communicative abilities and their empathy to form other trusting relationships, in which they receive attention and recognition. As they develop their talents, children's curiosity and motivation to learn drives them to seek out the learning experiences they need. Adults satisfy their need for self-development by using and perfecting their talents (a musical gift, for example). In order to achieve things and further their careers, employees make use of abilities such as understanding numbers and planning, and skills such as reading and writing. People use a huge number of abilities and skills to earn a living.

Satisfying our basic needs, then, depends on how our competencies are configured and how we use them. If someone has very good linguistic competencies, for instance, his need to develop and use them is greater than if they are only average. We will go into this in greater depth in Chapters 5 and 8.

We can't just claim the right to satisfy our own basic needs. We must afford the same right to our children, partners, colleagues – to all our fellow humans, by showing understanding for their individual profiles of needs, however they are configured. Parents may expect their son to get into an academically selective school. But what if his abilities and his need for self-development and academic achievement are too low to fulfil their expectations? Or in the world of work: are the demands a manager makes of her team in line with their individual basic needs and abilities? Or is she overstretching them because she wants to use their achievements to further her own career? Our fellow humans' individual profiles of basic needs and competencies pose many challenges for us, in our families, friendship groups and at work. They demand from us not only a high level of tolerance, but also the willingness to support others in their efforts to satisfy their needs. To modify Kant's categorical imperative, we should always remember: do not place your own basic needs above the basic needs of others. And satisfy your needs in a way that doesn't hinder other people's satisfaction of theirs.

A society oriented towards basic needs

If you deal with society you must accept its ways, for its ways are your ways. Your needs and demands have created them. Your desires are so complex and contradictory – no wonder the society you create is also complex and contradictory.

Nisargadatta Maharaj (1897–1981)[2]

Our ancestors were self-sufficient for over 200,000 years. Most of the time their lives were difficult and shaped by individual responsibility. In communities, they worked together to provide food and combat existential threats such as drought or enemy attacks by other tribal groups. At the start of the Industrial Revolution, self-sufficiency began to decline rapidly, and with it self-determination and self-reliance. After just a few decades, people could no longer produce all their own food. Today, hardly anyone is self-sufficient; in our increasingly industrial agricultural economy, not even farmers fall into this category. Our groceries are provided by a global food industry. In the space of 150 years we have said goodbye to a way of life that had existed since the emergence of *Homo sapiens*. But the need to shape our own lives independently – to some degree, at least – is still there. It is no surprise, then, that more and more people are feeling alienated and suffering from hard-to-define anxieties.

Today, social and economic institutions have taken over the job of satisfying some of our basic needs. A comprehensive health system looks after our physical integrity. A highly developed education system allows us to fully exploit our potential. Functioning security and legal systems protect us and our property from all forms of violence. But none of these anonymous institutions is capable of satisfying our need for emotional security or giving us a sense of belonging and social recognition. This can only be accomplished within a community of people who know each other well. These communities have become increasingly rare in modern society and their decline has had a negative effect on people's wellbeing. Reshaping society in such a way that, in future, more people can live in this kind of community is one of the Fit Principle's most important aims.

People increasingly feel like small cogs whose job is to keep the machine of state and economy running. Instead, society and the economy should serve people, by providing the framework in which they can satisfy their basic needs as independently as possible. In Chapter 10 we will look at how a balance can be achieved between the individual freedom to satisfy our basic needs and the reasonable demand that, despite our diverse range of needs and talents, we should create the fairest possible living conditions for all.

5

Competencies We Want to Develop

*Humans achieve countless things that no other
living creature is capable of.*

We humans are right to be proud of our abilities. We achieve countless
things that no other living creature is capable of. We construct cars
and planes and are constantly inventing new ways of communicating,
such as the internet or the smartphone. And yet we find it difficult to
pinpoint abilities that are specifically human and clearly separate us
from other living things. In fact, we see many similarities between
ourselves and a wide variety of animals. The primates exhibit sophisticated relationship and communication behaviour that is closely
related to our own, and, like us, they use objects as tools. We're not
the only ones to live in huge social systems: ants and bees do, too.
Harvest mice and weaver birds build impressive houses – even if there
is no variation in the design. And there are a lot of things animals can
do better than humans. A cheetah can run at more than double the
speed of Usain Bolt, the current 100-metre world record holder. Some
animals also have more highly developed senses than humans, like a
bird of prey's sight or a dog's sense of smell. Some species even have
information systems for which humans lack the necessary sense
organs. Bats and dolphins can send and receive ultrasonic signals,
enabling them to perceive the details of their physical environment
even in total darkness. Humans have no specific organ that makes
them unique. So what distinguishes human intelligence?

This is a question that has occupied me for more than thirty years.
I have observed children from birth to adolescence as they develop
their competencies. And I have learned that the best way to understand how abilities such as proficiency with numbers or ideas such as

morality are acquired is to look at the specific experiences children have in the course of their development – in play, for instance. In Chapter 5 we will look at how children develop their abilities and in so doing gradually arrive at intellectual ideas. This detailed account should help readers understand the elements that go to make up human intelligence and how it develops. In the process, we will show the crucial significance that competencies have for the satisfaction of our basic needs, and the extent to which they shape us and our individuality.

WHAT WE MEAN BY INTELLIGENCE

Questioning the world

> Man is a creature who searches for causes; in a hierarchy of minds he could be called a 'cause-seeker'. Other minds perhaps conceive of things according to different relations that to us are incomprehensible.
>
> Georg Christoph Lichtenberg (1742–99)[1]

Humans are intelligent creatures – but they certainly aren't the only ones. Over the course of evolution, intelligence in its broadest sense developed from an organism's need to spot characteristics and connections in its physical and social environment and react in a way that would benefit that organism. All living things – animals, plants and even microbes – are able to acquire the information they need for their survival. The gut bacteria *E. coli* can differentiate between nutrients and poisons. Receptors in its cell walls produce chemical signals that cause the bacterium to approach a positive stimulus or flee from a negative one, propelled by its whip-like flagella. In more advanced animals and humans, information is gathered by highly developed sense organs such as eyes and ears and processed in the brain. If the information is judged to be significant, it is stored away. If it appears to be useful to the organism in some way, the organism will react to it. Since information from the animate or inanimate environment has a different significance for humans and other animals, they have developed their own sense organs and brain functions to process it. All the higher animal species have organs for seeing and hearing, but the information that their brains process differs greatly from one species to another. This means there is no single intelligence. Each species has its own form of intelligence to suit its needs. These 'intelligences' are a product of evolutionary adaptation, through which each living thing only registers those aspects of its environment that have proven to be useful or harmful to it. It was only when *Homo sapiens* appeared on the scene that a species began to value knowledge for its own sake. We think about the universe without drawing any immediate use from the knowledge. We aren't separated from the more highly developed animals by some new and unique ability; the abilities we have in common with them are just more highly developed in humans.

The philosopher Georg Christoph Lichtenberg saw man as a 'cause-seeker'. In humans, animal learning behaviour has become a genuine need to understand the environment (see Chapter 1). A cat is interested in water, but only when it's thirsty. Once it has slaked its thirst, it loses interest. Not so with humans. Even toddlers are interested in water, whether they are thirsty or not. They turn taps on and off repeatedly, watching the water flood out and then stop. They pour water from one container into another and back again. They pour water over sand and watch the sand become hard as it seeps in. These games are how children discover the properties of water. In the same way, generations of chemists conducted countless experiments to discover that the water molecule consists of one oxygen atom and two hydrogen atoms. Humans want to make practical use of their knowledge, too. Engineers have constructed increasingly efficient turbines to create an electric current from hydraulic power, and over the centuries sculptors and landscape gardeners have created ever more beautiful fountains and lakes.

As the developmental psychologist Jean Piaget showed, knowledge and the things we make with it are all dependent on 'symbolic ideas', which originate in people's experiences of their physical and social environment. All the games children play with water create the symbolic idea of water. Symbolic ideas build the scaffolding of meaning required for language, which we call semantics. Our ancestors probably started using words out of a need to exchange symbolic ideas – about different types of animal, for instance – with other people. It is symbolic ideas that have made cultural, technological and economic progress possible, and made humans unique.

Of course, the more highly developed animals also have their concrete ideas – a cat has the idea of a mouse. But our symbolic ideas and the linguistic terms that go with them are much more comprehensive and have a special quality: removed from the experiences that first generated them, they can be used in new combinations. We can write a shopping list even when we're not hungry. In January, we can talk about the holidays we had last summer. We can discuss the Arab Spring or which country might win the next football World Cup. We connect linguistic terms and ideas in new ways time and again. If human language merely served to transmit signals, as animal communication does – a marmot standing guard whistles to warn its grazing family of the fox creeping towards them – a culture like ours could never have developed. A highly developed store of ideas and highly sophisticated language have given humans their inexhaustible creativity and immense productivity.

There is more to intelligence than an IQ test can show

The intelligence quotient is used as a measure of intelligence all over the world.[2] It assumes that a common intelligence factor plays a greater or lesser part in all cognitive achievements. And there are certainly functions such as short- and long-term memory, or the executive functions we use to plan future actions, that are of general importance for our cognitive performance.[3] But even the ancient world recognized that humans have a number of different mental abilities. In the Middle Ages, intelligence was divided into different categories: the *trivium* (grammar, logic and rhetoric) and the *quadrivium* (arithmetic, geometry, music and astronomy). Today we recognize a large number of intellectual abilities that can be distinguished by their functions and their underlying organic structures. If we are going to understand diversity and individuality, we must recognize that cognitive abilities are set at different levels in different people, and specific abilities are also set at different levels within one individual.

Aiming to do greater justice to the whole range of human intelligence,

Howard Gardner introduced the concept of multiple intelligences to psychology in 1983. He described six sub-forms of human intelligence, which he later modified and expanded to eight.[4] Gardner's theory received little recognition. However, a model of multiple intelligences such as the one used in this book represents individual ways of thinking and acting much better than a single number like the intelligence quotient.

This chapter introduces eight areas of human intelligence which – to prevent any misunderstandings – are termed 'competencies' rather than 'intelligences'.

People naturally think of competencies like language or logical reasoning as intelligence; social, motor and kinaesthetic competencies much less so. But they exhibit the same characteristics we ascribe to language and logic. Like all competencies, we use them to tackle the challenges life throws at us.

Each of the eight competencies is based on abilities we originally shared with the more highly developed animals, which we developed further over the course of evolution. Figural and spatial competencies, for example, come from visual and tactile-kinaesthetic perception.

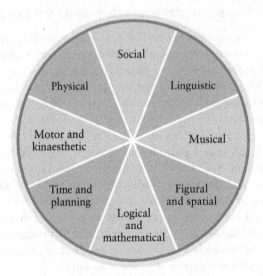

Figure 5.1: The eight human competencies.

A child's experiences with objects in the first years of life allows her to develop an understanding of properties like 'red', 'round' and 'bouncy'. The child brings these ideas together in the word 'ball'. Only then can she understand her mother when she tells her to 'fetch the red ball'. In the first years of life, our competencies are linked and woven together. The child is now capable of using her ideas creatively: she paints a red ball. Writing uses linguistic, figural and motor competencies. At school, the child learns to quantify her spatial ideas, by using her understanding of numbers. She measures the dimensions of objects and learns to calculate volume. And, as an adult, she might go on to build houses or design furniture.

An understanding of the competencies and how they work together gives us a greater understanding of how we and our fellow humans think and act. Because social and linguistic competencies are especially important to us, they are described in more detail here than the other competencies.

SOCIAL COMPETENCIES

A well-dressed, middle-aged woman is sitting alone in a train compartment. She is studying bank statements; a briefcase full of documents lies open on the seat opposite her. At the next stop an older gentleman gets on. He is neatly dressed and wearing a hat. He enters the compartment only after being invited with a friendly nod from the woman. The man chooses the seat furthest away from her.

The two travellers haven't exchanged a single word, and yet they have communicated with each other. If a younger man covered in tattoos and wearing a leather jacket and boots had barged into the compartment, the woman might well have given him a less friendly reception. The woman's age and clothing and the documents spread out in front of her also influenced the man's behaviour. He now notices the woman watching discreetly as he stows his luggage away and takes off his hat and coat. In the first few seconds of their encounter, they have both communicated something about who they are, and what they think and expect of each other. When they finally exchange greetings, it isn't just what they say, but their tone of voice that each of them finds significant.

Whenever people meet, there is always some form of communication. Even if they ignore each other, they have still created a relationship. Humans just can't help communicating.[5] We are deeply social creatures who need other humans for our mental wellbeing. Entering into relationships and conversations is a basic need, and this is the reason we communicate.

The huge number of psychology and self-help books available bear witness to our need for a better understanding of our own behaviour and other people's, to help us get along as well as possible in our dealings with them. This isn't always easy: our thoughts and feelings, motivations and behaviour are so deeply anchored in our unconscious and our approach to life that we often have trouble finding a rational way to think about our social competencies. No wonder that, in the areas of child-rearing, relationships with partners and dealings with colleagues, we very often follow our gut instinct.

In more recent literature, social competencies are frequently called 'emotional intelligence'[6] – a rather unfortunate expression, since social competencies are made up of much more than just emotions, as we will see. There are four essential building blocks that go to make up the social competencies, the combined effect of which has a huge significance for how we behave in relationships. They are: nonverbal communication (body language, the perception and expression of social signals); attachment and caring behaviour; imitative and social learning; and social cognition.

Nonverbal communication

The beginnings of nonverbal communication can be found a long way back in our evolutionary history. Many hundreds of millions of years ago, microorganisms exchanged information and reacted to one another with the help of chemical messengers. When the first animals appeared, around 400 million years ago, they developed simple forms of communication, such as the recognition of smells, sounds and shapes, from which they eventually built up a more nuanced body language. This helped the more highly developed animals regulate relationship behaviour within their species and their behaviour towards other species, communicating affection or aggression. Animals use body signals that humans understand as well. Cats arch their backs and lay their ears flat against their heads when they see a dog, in a threatening gesture. Dogs wag their tails to show sympathy, or bare their teeth to signal they are prepared to attack. The repertoire of human body language is very broad: we give off a host of social signals through facial expression, eye contact and tone of voice. The following catalogue shows just how wide our range of social signals is:

Facial expression. Joy, sadness, mistrust, surprise and fear – our faces express the whole palette of our emotions. Our mouths, the lines on our foreheads, our eyes and eyebrows and the angle of our heads all serve as a means of expression. Each part of the face acquires a specific meaning. Eyes have such a strong emotional effect on us that, for example, in Europe we can't help seeing a pair of eyes in two horizontal dots on a round electrical plug.

Eye contact. Two people look deep into each other's eyes for an unusually long time. They might be a mother talking to her baby, or lovers losing themselves in the unfathomable depths of the other's soul, or two angry people staring each other down. And if we look at the floor rather than into the eyes of the person we're talking to, it doesn't matter how convincing our words sound, that person will sense that we're not interested in them – or that we are so interested we don't dare look at them. If we look someone in the eye a moment too long, or not long enough, we communicate quite different information to them. Looks speak volumes.

Voice. Our voices can be warm and welcoming or icily cold, ingratiating or spiteful. When we talk to people, our meaning is often contained less in what we say than in how we say it. A politician wins over the public by the rousing way he gives a speech. Reading the same speech in a newspaper, you quickly realize that it is entirely without substance. A professor has some sensational research results to announce – but because his voice is so monotonous, the important information barely registers with his audience. If what someone is saying doesn't match the way they are saying it, we usually find the tone of voice more credible than the actual words. When spoken with affection, 'you little rascal' becomes a term of endearment.

Posture. When we're tired, we let our shoulders droop. But if we're full of vigour, our bodies are taut. When they're feeling aggressive, some men stick their elbows out to make them look bigger. In conversation, women tilt their heads to one side and smile to indicate their agreement. Our posture communicates our emotional state and our attitude towards other people. If we're particularly attached to someone, we turn to them both metaphorically and physically; we mirror their posture, perhaps by crossing our legs in the same way as the person we are talking to.

Movement. Like posture, the way we move our bodies also reflects our emotions. If we're impatient, we shuffle around on our chairs, swing our legs or fiddle with our clothes. Talented dancers can express a wealth of emotions through movement, from fizzing joy to deep melancholy. Soldiers on parade swing their arms and legs at certain strictly controlled angles to give an impression of discipline and latent strength.

Distance. Like all animals, humans have a well-developed sense of personal space. Each of us is surrounded by an invisible but well-defined buffer zone, which can be larger or smaller depending on the culture we live in and the people around us. The distance we want other people to respect is determined by the situation, our degree of intimacy with the other person and the norms of our specific culture. If a stranger barges into this zone, it triggers aggressive behaviour or a flight response. If we're lying on a deserted beach, we are irritated if a stranger settles down ten metres away. On a rush-hour train, by contrast, we're fine with another person standing cheek by jowl with us for an extended period of time. But we still keep the stranger at an emotional distance by avoiding all eye contact – no matter how long the journey is. We adapt our personal space intuitively dozens of times every day to fit different situations. The shop assistant, the bus driver, the manager, the favourite relative, the detested neighbour – the distance we keep from each of them fits the person and the situation.

We expand our expressive repertoire even further by combining the elements of body language. If we are really delighted by something we may widen our eyes, raise our eyebrows, let out a cry of happiness and maybe even do a little dance of joy. And when we're miserable, deep lines form on our foreheads and between our eyebrows. Our eyes are dull, our eyelids and the corners of our mouths droop. Our voices are quiet and husky. We hunch our shoulders, and our movements become slow and cautious.

We have a very highly developed ability to perceive even the subtlest of nuances in other people's body language.[7] And we don't just read the signals; we also take account of their strength – how vigorously and how long someone nods their head, for instance. The extent to which the signals correspond to our own emotional state and that of the person we are talking to is particularly significant. It's confusing if someone responds to a mild shake of your head with a furious tone of voice.

Which elements of nonverbal communication are innate, and which are learned? The ability to perceive social signals and to express them is innate, but the meaning of physical signals and how they are used is something children learn from their parents, attachment figures and peers. The signals we use – shaking hands when greeting someone, for example – are not instinctive; children internalize them through social interaction (see Chapter 3). Every culture ascribes its own significance to social signals and has its own conventions for using them. Children rely on extensive interpersonal experiences to help them acquire all the different culturally specific forms of expression used in nonverbal communication.

In the first two years of their lives, children can't communicate without the help of body language.[8] Spoken language only becomes important after that point – and nonverbal communication remains a crucial element of relationship behaviour throughout childhood and into adulthood. Even babies can read simple body signals, particularly facial expression and tone of voice, and they can express feelings like fear, disgust and curiosity. During childhood, children start to understand more nuanced nonverbal communication. In their 1994 study, the behavioural researchers Stephen Nowicki and Marshall P. Duke looked at how the understanding of facial expression develops between the ages of five and twenty. They showed children of various ages a series of twenty-seven pairs of pictures. Each pair showed two people with different expressions, such as one cheerful boy and one angry one. The children were asked which boy was cheerful and which was angry.

They discovered that the ability to interpret someone's facial expressions correctly increased steadily from nursery age to adolescence. Individual achievement varied at each age. There were some seven-year-olds who managed to interpret fewer than eight pairs correctly, and others who were able to categorize more than twenty, which is an above-average score for an adult. And there were young people who were no more competent at the age of fourteen than the average seven-year-old. Substantial gaps between individuals emerged in another study by the same researchers, which looked at the ability to interpret the emotions expressed in people's tone of voice – kindly or enraged, for example. All the signals involved in nonverbal communication are developed to different levels in different people. Some children and adults can perceive even the subtlest changes in the facial expression, posture or voice of the person they are talking to

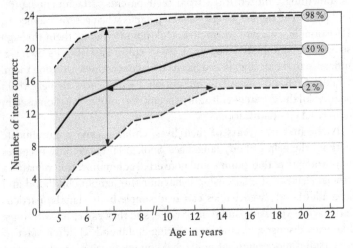

Figure 5.2: Recognizing facial expressions. Children were shown twenty-seven pairs of pictures of various expressions and asked to identify them. At the age of seven, the most competent children were able to correctly identify almost three times more pictures than the least competent. Fifty per cent of the seven-year-olds achieved a result that other children managed only at the age of fourteen, or not at all. Fifty per cent corresponds to the mean, 2 per cent the lower boundary of the normal range, and 98 per cent to the upper boundary. (source: Nowicki and Duke 1994)

and react to it appropriately. Others can't read social signals well enough, or can't really read them at all, and therefore have a tendency to behave inappropriately in social situations. A few people suffer from a very specific weakness in their visual perception: they are unable to recognize people by their faces. Children affected by this condition don't even recognize their mother's and father's faces (the condition is called 'face blindness' or prosopagnosia).[9] It is estimated that 1 per cent of the population suffers from face blindness, including the famous primate researcher Jane Goodall.[10] She once told an interviewer how uncomfortable she felt when she was unable to recognize someone coming towards her, even though she had known them for decades.

These signals vary not only from one person to another, but within individuals. This means that someone can have a very lively face, but a rather flat, monotonous voice. For other people, the opposite is true – or someone's facial expression and tone of voice might both be strong or weak.

Interestingly, the abilities involved in nonverbal communication are developed to a slightly higher level in women than they are in men.[11] Anthropologists put this difference down to women having evolved a higher sensitivity to nonverbal communication in order to empathize with the behaviour of babies and small children and respond appropriately to their needs. But the overall difference in nonverbal abilities between the sexes is much less than the difference between individuals of the same sex.

Attachment behaviour and caring

Over the past 40 million years, some animal species have developed pronounced attachment and sophisticated breeding behaviour. In mammals and birds, the bond between a mother and her young is produced by various hormonal and neuro-physiological mechanisms. 'Imprinting' is a widespread bonding mechanism in the animal kingdom.[12] Ducklings bond with the first living thing they see after hatching. If this happens to be a human rather than their mother, as nature intended, the ducklings will bond with that human. Cows that have just calved release the hormone oxytocin, motivating them to bond with their calf and care for it. These automatic processes play a more subordinate role in humans. The bond between a child and his parents or other attachment figures is largely formed through familiarization and shared experiences. Over the first few years of life, children develop a profound emotional and physical dependency, expressed in typical behaviours such as the desire for closeness and affection, fear of strangers and separation anxiety.[13]

A minimum level of emotional security and affection is a biological necessity for every human, but especially for children. Children cannot be alone. Their sense of wellbeing depends on love and affection from familiar people who can be relied upon to care for them. This need is as elementary as the need to satisfy hunger and thirst. A sustainable, two-way relationship between children and their parents, but also between children and other attachment figures, is vital if they are going to develop successfully, in a process that takes at least fifteen years. There are three important reasons for this:

Children are reliant on the care provided by their parents and other attachment figures. They need to be fed, looked after and protected. They wouldn't survive without this care.

Children need their parents, other attachment figures, siblings and other children as role models. This is the only way for them to acquire the multi-layered relationship behaviour unique to human beings.

Only when children feel emotionally secure and accepted can they fully develop their curiosity and motivation to learn, and have the experiences they need for their development. If they feel emotionally neglected, their development will be delayed (see Chapter 3).

But childhood bonding serves another important purpose: children obey their parents out of a natural respect for their authority that comes from emotional dependence. In the main, children are obedient (most of the time) because they are attached to their parents and carers and don't want to lose their attention and affection. Setting boundaries and giving out punishments is still necessary, but a child's level of obedience depends much more on how trusting her relationship with her parents and carers is. Parents whose children exhibit autistic behaviour know how difficult child-rearing can be. These children feel less attached to their parents, which means bringing them up can be a very complex business.

Human children function according to the same biological behavioural principles as the young of birds and mammals.[14] Children remain bonded to their primary carer – as a rule, their parents, and in particular their mothers – until they can survive on their own. When they have matured physically and can look after themselves, the bond with their parents loosens enough for them to become emotionally and socially independent. From the behavioural biology point of view, that bond has served its purpose. This doesn't mean that adults lose their willingness to bond; they simply do so much less than children.

Children don't just bond with their biological parents: they form attachments to anyone who cares for them, protects them and gains their trust. The child psychiatrist John Bowlby talks about 'instinctive' attachment behaviour.[15] Another important thing to note is that the bond children form with their parents is unconditional. It exists whether a child's parents love and care for her or are cruel and uncaring. No matter how much children's parents or carers neglect them, they will never fundamentally question the relationship or break it off altogether and go in search of other parents. Children are unconditionally devoted to their parents and are therefore entirely at their mercy.

Children learn to trust the people who care for them adequately. The strength of their bond with parents and other attachment figures therefore depends essentially on the time they spend with them. But children don't just want to bond: they want to feel happy and comfortable. The quality of the parent–child relationship is the crucial factor in children's wellbeing. The way in which parents and other carers satisfy children's needs, in particular the need for emotional security and affection, is the decisive element in their mental and physical wellbeing. In order to engage with children sensitively, an attachment figure must be familiar to the child. She needs to be available and reliable, and to react appropriately to the child's individual needs. If the attachment figure fulfils these conditions, she will become a bastion of emotional security and physical wellbeing for the child.

Attachment behaviour always follows the same course in childhood. The bond grows rapidly after birth and reaches a peak in the first years of life, weakening slowly but steadily from that point as the

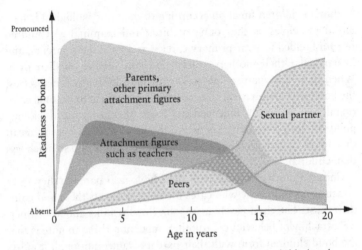

Figure 5.3: The development of attachment behaviour. A child's bond with her parents and carers is strongest in the early years of life, before weakening gradually, and then more dramatically during adolescence. In its place, the willingness to bond with peers, and eventually with a partner, increases. The shaded areas show how much the willingness to bond can vary between individuals.

child heads towards puberty. Children also bond with other attachment figures such as grandparents, childminders and teachers, though to a lesser degree than with their parents. They become attached to their siblings and other children who become their friends. In adolescence, the bond with their parents loosens enough for young adults to be able to leave the family. They don't become emotionally independent, however; they seek emotional security from their peers and ultimately form a bond with a partner. Falling in love is similar in its intensity to the early childhood bond, though regrettably it doesn't last as long. As adults, most people retain a certain degree of intimacy and trust with their parents, though it doesn't restrict their emotional independence.

As the graph shows, children bond to different degrees. Some have a great need for intimacy and affection. We regard children who are reluctant to be parted from their attachment figure and are very frightened of strangers as shy. At the other end of the scale, there are children who quickly develop a degree of emotional independence from their parents. They willingly enter into relationships with other people even as babies and toddlers, and bond with neighbours and nursery workers. The willingness to make friends with children of their own age differs from one child to the next as well. Some have just one very close friend, while others have a number of more casual friendships.

A parent's bond with their children is not as unconditional as children's with their parents, but it is still strong. When a baby is born, his parents are strongly motivated to accept and care for him. Over the years, they may go to great lengths and make huge sacrifices for his sake, if they have to. Parents don't just bond with their children; they have a genuine willingness to care for them. Children stimulate their parents' desire to give them attention and affection in various ways. Babies and small children have certain physical features that trigger affectionate behaviour in adults. Konrad Lorenz called these features the *Kindchenschema*, or 'small child pattern': in comparison to adults, children have proportionally larger heads and smaller bodies. They have large foreheads and small faces. Their cheeks look round and full, and in those small faces, their eyes often look enormous. Young mammals (kittens, for example) have the same features. Looking at them prompts feelings similar to those we have for children. The motor function seen in small children is something else that reinforces our desire to care for them. A baby lying helplessly on his back, kicking his legs in the air, triggers our protective instinct. A child reaching his little arms out to us is irresistible: we have to cuddle him. Nature has found all kinds of ways to make children appealing.

Parents take care of their children's mental and physical wellbeing; they bring them up and support them in their development. In many respects this is a huge achievement, and it takes at least fifteen years. But it isn't all about giving: parents are also richly rewarded by their children. When babies are full and contented, when they are happy, smiling, burbling and squealing, their parents feel a deep sense of satisfaction and happiness. Just as they do when their children play and chatter to themselves, run around and cry out with delight. And they are proud when a child shows an interest in letters and numbers for the first time. That doesn't just apply to parents, either: grandparents, childcare professionals and teachers all feel drawn to the children they care for, and experience tremendous satisfaction when they develop well.

Like children's willingness to bond, caring behaviour in parents and attachment figures is set at different levels from one person to the next, and is slightly more pronounced in women than in men. But there are plenty of fathers who take great pleasure in looking after their children and do so with tremendous dedication. Our caring behaviour benefits not only children but people of all ages, in particular the sick, the disabled and older people.

Imitative and social learning

Once upon a time there was an old, old man. His eyes had grown dim, and his ears deaf, and his knees trembled. When he sat at the table, he could scarcely hold a spoon. He spilled soup on the tablecloth and dribbled some of it from his mouth. His son and his son's wife were disgusted by this, and eventually they made the grandfather sit in the corner behind the stove, where they gave him his food in an earthenware bowl. He sat there looking sadly at the table, and his eyes grew moist. One day his shaking hands could not hold the bowl, and it fell to the ground and broke. The young woman scolded, but all he did was sigh. Then they bought him a wooden bowl for a few hellers and made him eat from that.

One day when the parents were sitting at the table, the little grandson started to push some pieces of wood together on the floor.

'What are you making?' asked his father.

'I'm making a little trough for you and mother to eat from when I'm big.'

The man and his wife looked at each other and then began to cry. At once, they brought the old grandfather to the table, and from then on they always let him eat with them. And if he spilled a little, they didn't say a thing.

The Brothers Grimm, 'The Old Grandfather and the Grandson'

No matter what culture they live in, all parents and carers have the same goal: their children need to learn how people treat each other in their community. Children have to follow the rules of interpersonal relationships and take on the community's values and customs.

Children are socialized much less by being told how to behave than by learning from role models – and this is thanks to our innate talent for imitative or social learning.[16] The story of the old grandfather and his grandson sheds light on how the process works: the child takes his cues from the people he lives with, mimics their behaviour and internalizes their values. A child's interpersonal behaviour and ideas about morality are taken from his role models, and he cannot be any different to them. The cabaret artist Karl Valentin put it very well: 'We can't educate children; they copy everything we do.' Some anthropologists believe that imitative learning has played a fundamental part in human social evolution,[17] enabling patterns of behaviour and values to be passed down from one generation to the next.

'Mirror neurons' play an important role in imitative learning. In 1992 the neurophysiologist Giacomo Rizzolatti and his team measured the brainwaves of one group of macaques as they ate nuts, and a second group who were watching the others eat through a glass window. The scientists found that the same area of the brain was active in all the monkeys, no matter whether they were carrying out the actions themselves or just watching.[18] Numerous other research groups have now studied the phenomenon of mirror neurons. It has become increasingly clear that the networks in which these neurons are involved are capable of much more than just perceiving things or performing actions. They also enable us to experience things vicariously and imitate other people's behaviour. These networks are closely connected with emotions and somatic-vegetative functions, which means that when we watch someone else eating, for example, we start to feel hungry as well.

Children are born with a desire to mimic and internalize other people's behaviour and values. Even newborns can imitate the facial expression of someone who opens their mouth and sticks out their tongue.[19] The development of 'symbolic play' is a very good demonstration of how children internalize patterns of behaviour and actions.[20] At the age of twelve to fifteen months, children begin to acquire the ability to use objects functionally. A child will try to use a spoon herself when her mother is feeding her (direct imitation). In the hours and days that follow, she will play out the situation again and again, bringing an empty spoon up to her mouth (functional play). At the age of 15–18 months, that game will enable her to develop her first symbolic ideas.[21] These ideas are independent of a specific time and place, and the child can therefore apply them to all kinds of different situations. She doesn't just perform an action such as 'eating from a spoon' on herself; she feeds other people or dolls (representative play I). She then takes another step by imagining that the doll is using the spoon itself (representative play II). By the time the child is three, her powers of imagination are sufficiently developed for her to re-create not just single actions, but whole sequences, such as a family meal at the dinner table (sequential play). Children don't imitate any and all behaviour, just the behaviour that fits their level of development. By the age of three or four, their fine motor skills have developed enough for them to produce scribbles as they try and imitate their parents' and older siblings' writing. Once they have started school, the need to belong to a community leads them to adopt social rules, values and roles – imitating a friend they admire, for example. The older they get, the more important children their own age and slightly older become as role models. There are many things that children learn better and more quickly from other children than they do from adults. In adolescence, their peers become their most crucial role models, more important than parents and other adults.

In the first years of life, children chiefly acquire interpersonal behaviour from the people they know well. A high level of emotional dependence on their parents makes them very willing to take their cues from them. In their roleplays, children delight in imitating people who have made a particular impression on them, such as the carer at nursery or the GP during their most recent visit to the doctor. Children also like to adopt behaviour that adults and other children seem to enjoy and carry out successfully. Ultimately, children's choice of role models is also determined by their stage of development and their individual abilities. Their choice tends to reflect where their own talents lie; sporty children might pick a famous footballer or a successful tennis player.

Our willingness to copy role models is particularly strong in childhood, although it is different for every child. Some children concentrate to an almost worrying degree on one particular person – a pop star, perhaps, or even a fictional character such as Superman. Adults have role models, too: we might imitate someone else's hairstyle or fashion sense, or adopt the views of politicians we agree with. Anyone taking up a leadership role in schools, society or industry should be aware of their role model function and the responsibility that comes with it.

Social cognition

> Confucianism, fifth century BC: What you do not want done to yourself, do not do to others.
>
> Confucius (551–479 BC)[22]

> Hinduism, fourth century BC: One should not behave towards others in a way that is disagreeable to oneself. This is the essence of morality. All other activities are rooted in selfish desires.
>
> Mahabharata[23]

> Judaism, second century AD: That which is hateful to you do not do to another; that is the entire Torah, and the rest is its interpretation.
>
> Talmud, Shabbat 31a

> Christianity, sixteenth century (Lutheranism): See thou never do to another what thou wouldst hate to have done to thee by another.
>
> Tobit, 4:16, Apocrypha

The Golden Rule appears in all the major cultures and religions of the past 2,500 years. At its root lies an extraordinary ability: humans are able to empathize with the feelings, thoughts and actions of others, and judge what effect their own thoughts and actions will have on them. Ideas like the Golden Rule are part of social cognition, which developed from our elementary social competencies over the course of evolution.

In child development, the beginnings of social cognition can be found very early on, in the first forms of relationship behaviour. Babies learn, albeit unconsciously, how their behaviour affects their parents, and what reactions it elicits. During childhood, countless interactions with other people help children understand the rules of relationship behaviour and ultimately the values and morals that a community lives by.

Introspection and extrospection

The first significant development in terms of social cognition – the first time children extend their consciousness – happens between the ages of 15 and 24 months, when they start to see themselves as people.

A reliable indicator of whether children have reached this level of self-perception is known as the 'rouge test' because red was the colour originally used for it. When children begin to recognize themselves, they become aware that the spot of colour doesn't belong on their face.[24] The test was first performed on apes.[25] Chimpanzees and orangutans noticed the spot on their faces, but gorillas and other primates didn't. The test also revealed that chimpanzees raised in isolation didn't develop a sense of self. The sense of self is linked to social experiences, and it can only develop in a community.

As young children grow, they develop a fuller understanding of themselves as independent people and are able to separate themselves

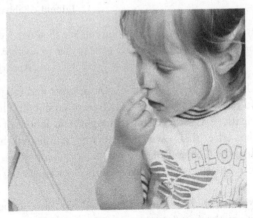

Figure 5.4: Rouge test. Stephanie, 24 months old, has a spot of colour surreptitiously painted on to her nose. She is then placed in front of a mirror. Stephanie notices the colour and puts her hand to it. (source: Zurich Longitudinal Studies)

from others – a fact expressed in language. They start to use 'I', 'you' and 'we'. Two words that pre-school children love to use are an expression of this emerging identity: 'I want'. At first, children assume that everyone thinks and feels the same things about the world that they do (a phenomenon Piaget termed egocentrism). At around the age of four they take another important developmental step, when they start to empathize with other people's emotions and understand their thoughts and how they think. In psychology, this is called the 'theory of mind'.[26] Children learn to understand others' emotions, behaviour and actions, and separate them from their own thoughts and feelings. The ability to think about our own state of mind (introspection), as well as that of other people (extrospection) is an essential prerequisite for the development of nuanced relationship behaviour.

Once they have been at school for a few years, children develop an 'interior language'. Their awareness of their own thoughts and feelings prior to this point has been patchy, but from then on they live in a constant stream of conscious experience, aware of their own thoughts, feelings and actions, and also able to question them.[27] They start talking to older children and adults about their ideas, and acquire new ways of seeing things.

The degree to which social cognition is refined over the course of childhood is shown particularly well in the ability to recognize jokes and irony. In the first years of life, children believe everything they're told. They are also incapable of lying – of deliberately saying something they know to be untrue. They have to understand that statements can be false before they can spot a lie or deliberately deceive others. This understanding is acquired at the age of five at the earliest, when children start to realize what lying means.[28] They also understand what it means to be honest – but they still take wit and irony at face value. Only once they are six do children begin to develop an understanding of jokes, and quite often it takes years for them to understand irony.[29] Grasping the subtle difference between jokes on the one hand, and lies on the other, is an extraordinary mental achievement; the difference consists solely of the listener being surprised *but not deceived* by a punchline when someone is being funny. This requires a kind of meta-understanding of the circumstances under which a statement should be understood as true or false by the person who hears it.

Children develop this understanding at different speeds, and even as adults some people's level of understanding is higher than others.

The ability to empathize with other people's thoughts and feelings, their intentions and expectations, is innate. But the ways in which children use this ability are acquired. Childhood experiences shape how we use introspection and extrospection, and which patterns of behaviour and value judgements we adopt. If parents and teachers treat children with sensitivity, respecting their thoughts and feelings, those children will use their empathy to treat other people in exactly the same way. But if a child's feelings are disregarded, her thoughts dismissed and her requests and desires ignored, she will encounter other children and adults with mistrust. Depending on the example children are set when it comes to behaviour and values, they will go on to treat others with empathy or attempt to exploit and manipulate them.

How morality develops

Moral ideas are always the expression of a cultural reality, and can never become the universal truths we would like them to be. The developmental psychologist Helen Keller[30] has pointed out that in Western societies, which are dominated by an individualistic approach to life, moral ideas are different from those in the collectivist societies of Asian countries. All the same, psychologists and educators have carried out numerous studies in the attempt to find a common foundation for humanity's range of moral ideas. Lawrence Kohlberg made a detailed study of moral development and described a sequence of developmental stages that children go through no matter what culture they grow up in.[31] In the first years of life, children learn that certain types of behaviour are desired and others are not; some are even forbidden and result in punishment – such as putting your fingers in an electric socket (obedience and punishment). At four years old, when children become capable of empathizing with other people, they learn an important lesson: you will treat me the same way I have treated you. If I hit you, you won't want to play with me any more. Children realize that the way they treat other people will have an effect on them in return (mutual accord). Between the ages of four and six, these experiences of obedience and punishment, and of mutual accord, give rise to children's first ideas about good (what you should do) and evil (what you must not do). They start to develop a strong interest in fairy tales – stories of human behaviour, with endless permutations of good and evil. According to Kohlberg, at pre-school age children's morality is based on authority, and centres on obedience, praise and punishment. Over the years that follow, children's black-and-white understanding of what is permitted and what is forbidden becomes more nuanced as they start to include human character traits, different ways of behaving and different situations in their thinking.

Once children have started school, their morality is increasingly shaped by their dealings with other children. They take on the rules and rituals used in their group (rules of the community). If someone wants to play with them, they will have to follow the group's rules. Children fall into line in an effort to gain social acceptance and approval of their actions. Parents and teachers use their authority to make children aware they have a duty to fulfil. They need to learn what orderly behaviour means: throwing litter in the bin rather than dropping it in the playground, for example. From the age of ten, children start to internalize the rules of the society they live in (law and order). These rules give them a gradual understanding of social structures and state institutions. The police force guarantees protection from violence, and the courts ensure that justice is done. Children form a set of moral principles that depend on their own ideas conforming to the prevailing norms, rights and obligations in their culture.

In adolescence, children develop a generally applicable morality. Ethical values such as freedom and justice shouldn't just apply to their own community or even the whole of society, but to all people and ideally to all living things in the world (universal ethics). Young adults have always been particularly receptive to this kind of absolute morality. Goethe's hypersensitive hero Werther takes extreme care not to tread on any insects that cross his path while he is out walking. Today, the younger generation is much more active in the fight against poverty; they campaign for peace and against war, for the preservation of animal and plant species, and against climate change. They adopt a morality based on conscience and universally valid ethical

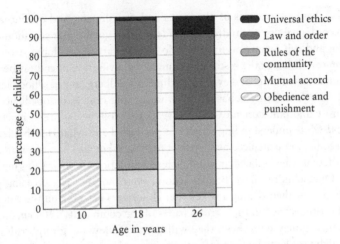

Figure 5.5: Moral development. Obedience and punishment: children are guided by adults' rules. Mutual accord: children realize their behaviour influences others and also has repercussions. Rules of the community: children understand simple rules of behaviour in the community. Law and order: young people abide by the rules of the society they live in. Universal ethics: young adults advocate an ethics that should apply to the entire human race. (source: Kohlberg 1976)

principles, which includes all humans and the greatest possible number of other living things. Different moral ideas prevail at every age. Ethical awareness is developed to different levels in young people and adults. Only a small minority believe in a universal ethics. Most people live by a morality of law and order that is valid for the society they live in – and some just obey the rules that regulate the interpersonal behaviour in their own small community (clan behaviour). And a huge range of moral ideas are also reflected in divergent political beliefs.

Over the years, various people have challenged Kohlberg's model of moral development. The psychologist Carol Gilligan was rightly disturbed by the fact that, in all the academic studies using Kohlberg's step model, boys' test scores were significantly higher than scores for girls.[32] She put the fact that girls seemed to lag behind down to the test's design. The conflict was sparked by test problems like the 'Heinz dilemma':

Heinz's wife is near death. The only pharmacist in the town has developed a drug that might save her. But he will only sell the drug at ten times the price it cost him to make it. Despite many efforts, Heinz can't get together enough money to buy the drug. In desperation, Heinz breaks into the pharmacy and steals the drug for his wife.

In the studies, boys tended to justify the theft of the medicine by saying that it would allow Heinz to save his wife. Girls argued that he should appeal to the pharmacist's sense of responsibility and try to change his mind, or go out collecting money from the people in his neighbourhood. Where Kohlberg's morality rests first and foremost on the principle of justice, Gilligan proposed an 'ethics of care'. Gilligan's morality has its roots in the family and is directed towards the needs of children, but is also intended to benefit all weak, sick and socially disadvantaged members of society, all over the world. Both ideas reflect different aspects of morality, from the family to the whole of humanity. The morality of justice (Don't do anything bad!) rests chiefly on instructions and prohibitions that are intended to regulate people's dealings with one another. A classic example is the Ten Commandments. John Rawls describes a modern form of justice in his standard work *A Theory of Justice*. He presents justice as a virtue that should be practised in the institutions of society and industry in such a way that the freedom of the individual is not restricted.[33] This form of morality seems to correspond to a more masculine way of thinking. By contrast, the morality of care (Do good things!) begins with caring behaviour that focuses on the needs of others. It is primarily, though not exclusively, women who live by this moral code. The Golden Rule becomes a morality of justice *and*

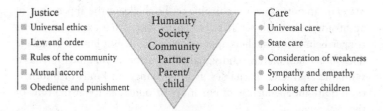

Justice	Care
▨ Universal ethics	● Universal care
▨ Law and order	● State care
▨ Rules of the community	● Consideration of weakness
▨ Mutual accord	● Sympathy and empathy
▨ Obedience and punishment	● Looking after children

Humanity
Society
Community
Partner
Parent/child

Figure 5.6: Moralities of justice and care.

care when we adapt it to read: 'Do unto others as you would have them do unto you.'

LINGUISTIC COMPETENCIES

> The jet ascends and, falling, fills
> The rounded marble basin up,
> Which shrouds itself before it spills
> Into a second basin's cup;
> Growing too full, the second runs
> Its surging billows to the next,
> And all three give and get at once,
> And run and rest.

Conrad Ferdinand Meyer (1825–98)[34]

Conrad Ferdinand Meyer's poem (especially in the original German) unites all the elements of language to great effect. The choice of words, their rhythm and melody, and the construction of the poem create an image in our minds and the emotions that go with it. As a deeply human phenomenon, language has always held a tremendous fascination for people of all cultures. We even believe we understand ourselves better when we understand the nature of language.

Human language is much more than just a means of communication. Animals communicate too, without using language. Blackbirds sing to attract a mate, and the chirping of their chicks prompts them to go and find food. They squawk loudly to ward off other birds trying to invade their territory. Communication among animals consists of signals prompted either by an internal stimulus (the need to attract a mate) or an outer one (the chirping of the chicks or the appearance of an invader). Animals send each other messages that express needs or trigger behaviours such as aggression or flight. But human language does more than merely send signals. Linguistic communication has a symbolic character. It enables us to exchange all kinds of information, quite independently of our inner or outer reality at the time of the exchange.

Our closest relatives, the great apes, occupy a special position in the animal kingdom. It has long been known that they are capable of making limited use of symbolic language. Chimpanzees can understand and use sign language.[35] But their linguistic talents remain modest, no matter how much encouragement they are given; at best, their understanding of language and their ability to express themselves remains at the level of a two-year-old child.[36] Recent studies, however, prove that chimpanzees living in the wild communicate using up to sixty-six species-specific gestures.[37] It seems the beginnings of human language may lie further back than we previously thought.

The elements of language

Human hearing is highly specialized, and our brains have two extraordinarily well-developed speech centres. One of these centres is used for understanding language; situated in the temporal lobe of the brain, it is called the Wernicke area after the man who discovered it. Spoken language is created in the other centre. It is found in the frontal lobe and known as the Broca area, again after the man who discovered it. These two very specialized language centres and their networks, which branch out and connect to other areas of the brain, are a clear indication that human language must have developed a very long time ago (see Chapter 1).

The exchange of symbolic ideas relies on a sophisticated set of linguistic rules. Sounds are combined to form words, which are tied to symbolic ideas such as 'child'. There are various ways to make individual words express more, such as forming singular and plural versions: 'child' becomes 'children' (grammar). Words are sorted into categories such as nouns and verbs. We also have specific rules for the order in which we use the words in these categories, for example, subject–verb–object, as in: 'child loves teddy' (syntax). These formal elements are what gives language its great expressive power. In the 1960s, the American linguist Noam Chomsky came up with the concept of language's 'deep structure'.[38] He argued that children are born with a predisposition for understanding the structure of language. To put it another way: children have an innate feeling for the rules of language and an ability to use them.

There are currently still around 3,000 languages being spoken in

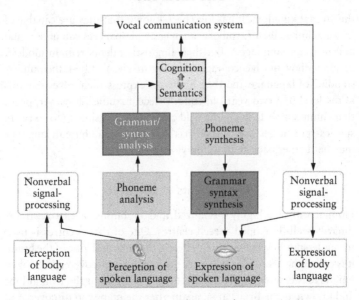

Figure 5.7: Human communication. White: social communication system, pragmatism; light grey: phonology; dark grey: grammar and syntax; bold outline: semantics and lexis.

the world. All of them rest on the same neurological foundation, which is made up of the following elements:

Perception and formation of speech sounds (phonology). Our brains process speech sounds in a completely different way from musical notes and noises (categorical perception).[39] The difference explains why our reaction to speech sounds also differs from our reaction to notes and noises. All people analyse and form speech sounds according to the same phonological rules. The linguist Leigh Lisker and his team looked at eleven very different languages, including Italian, German and Finnish, and discovered that the perception and production of sounds was based on the same phonetic regularities in all languages.[40]

Rhythm, volume and pitch (prosody). Prosody gives language its emotional quality and social meaning. A mother knows intuitively that her baby will pay attention and listen more if her voice is high-pitched and she speaks slowly, varying the pitch and giving her words a sing-song quality. She can increase the effect on the child by repeating what she has said many times over, with small variations in the words, pitch and speed.

Understanding and forming words and sentences (grammar and syntax). Individual words are formed and placed in a sentence according to strict rules that enable us to understand their symbolic content.

In the first years of life, children who are in the process of acquiring language achieve things that adults can very seldom do. Even a simple sentence such as 'Tom feeds the cat' poses incredible challenges to small children. They have to separate the sounds they hear into words and differentiate between at least two classes of words: nouns like 'Tom' and 'cat', which relate to an animal and a person, and verbs like 'feed', which relate to an action. Children understand that the form a verb takes is determined by a subject (Tom) and relates to an object (the cat). Children discover all of this for themselves. It is something no parent would ever think of explaining: 'Look, a simple sentence is made up of a subject, a verb and an object.' But children are capable of much more: they can learn to use singular and plural forms, to decline and conjugate, and to use the various time forms of verbs.

Children don't acquire language through imitation. If that were the case, then they would only ever be able to use sentences they had heard at least once before. The speed with which children grasp the many and various rules of word and sentence formation is a small miracle. And remember: they do it all on their own and without ever being conscious of those rules!

Meaning and vocabulary (semantics and lexis). Knowing the rules of spoken language doesn't mean that children are competent language users. For that, they need to grasp the meaning of words and sentences. They need to understand what has been said to them and express their own thoughts. If a mother says to her three-year-old son: 'Your car is under your bed', he doesn't just have to understand which objects, out of all the things in the house, are denoted by the words 'car' and 'bed'; he must know the meaning of 'your' and 'under' as well. This presupposes an understanding of personal attribution and an ability to imagine space. Otherwise the child will be unable to grasp which bed is meant and what spatial relationship exists between the car and the bed: the car isn't under any old bed, but under his; and it is *under* the bed, not on, behind or next to it.[41]

A word like 'jug' tells us how precise the meaning of words can be. A jug is an object that can be various shapes and sizes, but must have specific properties. A bottle is not a jug because it has no handle, and a cup isn't a jug either, because it has no spout. Another feature of words is that they can be put together with other words and used in endless new combinations – for example, to produce a figurative meaning: 'Little pitchers have big ears.'

Use of language (pragmatics). We use language differently depending on who we're talking to and the situation we are in. If we're talking to a small child, we use fewer words and simpler sentences than we might with a school-aged child or an adult. We also adapt our vocabulary and increase our use of nonverbal communication, such as facial expressions.

Small talk, the high art of socializing, is where pragmatics really comes into its own. The point of small talk isn't really to exchange information, engage in serious discussion or explore topics in depth. Small talk is exclusively a tool for forming relationships, showing the person you're talking to that you are interested in contact. The aim is to bridge the divide between people and become closer socially and emotionally. In short, to enjoy spending time together.

Our day-to-day language is a long way from being as elaborate as Conrad Ferdinand Meyer's poem. But we can still be proud of our linguistic competencies. Appreciating the subtleties of his poem is still a sign of an impressive linguistic ability.

How a child comes to language

Over the course of evolution, human language developed out of non-verbal communication (body language) in combination with our cognitive abilities and relationship behaviour. This process repeats itself in every child's development. In the first two years of life, parents and children communicate almost exclusively through body language. When a mother talks to her baby, the meaning of her words has very little significance for the child. The mother pays much more attention to the pitch, the melody and the expression in her voice. The baby shares his state of mind with his mother through facial expressions, eye movements and crying. Spoken language develops gradually, and is embedded in body language.

Children are little geniuses when it comes to learning. They can acquire any language spoken anywhere in the world. They 'listen in' to the language of their social environment and adapt the sounds they make to what they've heard. They pick words out of the long sequences of sounds they hear and gradually come to understand their meaning. In the first years of life, children learn several new words every day. By the age of two, they are building two-word sentences, and, at three to four years old, sentences of several words. At the age of five, most children can express themselves in whole sentences. By this point they understand around 4,000 words.[42]

The highly demanding process of language acquisition can only succeed if a child has constant linguistic interaction with parents, carers and above all other children. Nor is it enough to just hear language. The experience of language that children need is specific, comprehensive and contextualized. Only when children can make an immediate connection between words and the people, objects, actions, emotions and situations they refer to do they learn to understand language and ultimately to speak it. Language must therefore be embedded into children's day-to-day life and related directly to their experiences. Parents don't have to 'teach' their children language. They just have to enable them to have meaningful experiences with language. Unlike adults, most children can also learn a second or third language perfectly through the same process, though with a slight delay. The way children learn languages is known as synthetic language acquisition.

The ability to learn a language synthetically reaches a peak in the first years of life. It decreases steadily once a child has started school, and more or less comes to a standstill during adolescence.[43] Most adults are unable to learn a language holistically in the way children do; they have to learn words by heart and painstakingly acquire that language's formal elements (such as grammar and syntax). Usually, this type of language acquisition, known as 'analytic', will only lead to a limited competence, and the adult learner will always speak it with a foreign accent. There are some adults who have retained the

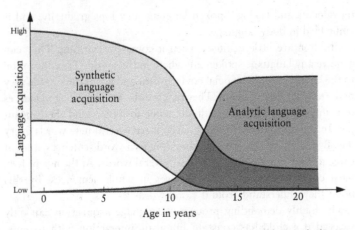

Figure 5.8: Synthetic and analytic language acquisition. The graph shows how we acquire languages, and the range of ability within a population. Synthetic: children acquire a language unconsciously, by connecting what they hear with nonverbal signals and experiences involving people, objects and actions. Analytic: adults learn a language consciously, by memorizing vocabulary and formal elements such as grammar and syntax.

ability to learn a language simply by talking to people, without taking lessons. But they are enviable exceptions to the rule.

Before the age of ten or twelve, the rules of grammar and syntax are a closed book to most children. It's only when they reach puberty and become capable of abstract thought that they are able to understand the rules of language and acquire a language analytically.

Before children can start learning written language, they need to have done much more than simply master the basic functions of language. Learning to read also involves figural and spatial competencies, which enable children to recognize individual letters and tell them apart – for example, the letters *b* and *d*. And writing also requires well-developed fine motor skills.

Diverse linguistic competencies

The ability to acquire a language synthetically is set at very different levels from one child to the next. The same applies to analytic language learning in teenagers and adults. And both language as a whole and its individual elements (such as articulation and semantics) can vary within a single person. Someone with a stutter might still express themselves very well in terms of meaning.

Differences in linguistic development quickly become apparent, even in the first year of life. Some children speak their first words at 10–12 months, and others don't speak until 28–32 months. Some manage to form two-word sentences at the age of 15–24 months, while others first do this at 30–42 months. At the age of six, children's vocabularies range from 1500 to 7000 words. Some are already speaking with the sophistication of an average eight-year-old, while others can hardly utter a sentence without making a mistake. Like speech, the development of written language diverges increasingly once children have started school. At the age of fourteen or fifteen, some students' reading and writing skills go far beyond the average performance of adults, while others have not got beyond higher primary-school level. It therefore comes as no surprise that there is a wide range of ability in written language among adults. Some have a perfect command of several languages, while others still only have a limited level of competency in their first language.

As with social competencies, there is also a small difference between the sexes in this area. Girls have a slightly higher level of linguistic competency than boys. But the average difference between the sexes is insignificant compared to the difference between the best and weakest boy, or the best and weakest girl.

Language has always been hugely important in all cultures. But no age has been so heavily dominated by communication as our own. The media and the electronic exchange of information permeate both our private and professional lives: in Western societies, about 70 per cent of people work in the service sector, and their jobs are largely based on an exchange of information. In our modern society, the way our communicative and social competencies develop largely determines the options that each of us has for self-development. These competencies have a huge influence on our lives.

MUSICAL COMPETENCIES

Music expresses what cannot be said and what cannot be left unsaid.

Victor Hugo (1802–85)[44]

A mother's singing calms her baby; a brass band gives soldiers their marching rhythm; a pop star sparks wild enthusiasm in teenagers; the church choir brings peace to the souls of the faithful. The power and the peculiarity of music lie in the fact that it is often better than the spoken word, or any other form of communication, at triggering emotions, moods and hidden longings. It is a catalyst for our emotional life, and gives us a strong desire to belong.

We don't know when our ancestors began to sing, dance and make music. The earliest musical instruments found by archaeologists are around 40,000 years old (see Chapter 1). But our forefathers were probably singing and dancing well over 100,000 years ago. There is a lot of evidence to suggest that musical competencies – much like the melodic and rhythmical aspects of speech (prosody) – developed from elements of nonverbal communication.

One indication that musical competencies such as singing and dancing developed very early on in human history is to be found in child development. Even in the womb, the foetus shows a preference for melodic and rhythmic speech.[45] At the age of two months, babies can recognize their mothers' singing by its pitch, volume and melody.[46] At a year old, children understand the rhythmic structure of songs, and in their second year of life they begin to join in and sing simple sequences of notes. Starting in their third year, children come to love rhymes like 'Wind the Bobbin Up', which are accompanied by rhythmical movements. At pre-school age children love singing songs together, and once they start school they are often keen to learn a musical instrument like the flute or piano. Children like to dance at every age, but particularly as teenagers. Dancing and listening to music together is an important component of youth culture: music brings people together and helps to break down barriers. Most adults listen to music and occasionally go to concerts. But their preferences for a particular type of music such as classical, pop or folk music are very different. Only a minority of adults sing or play an instrument.

Hardly any research has been done on our general levels of music appreciation and our ability to produce music. But it is clear that our sensitivity to rhythm, melody, harmony, dissonance and the timbre of notes differs greatly from one person to another. At the age of fourteen, the musical genius Wolfgang Amadeus Mozart heard the nine-part *Miserere* by Gregorio Allegri and managed to write out the score (which had been kept secret by the Vatican) from memory without a single mistake. A lot of people can only manage to sing a simple children's song from memory once they've heard it several times, and even then they will still make mistakes. But having a fine ear for music doesn't necessarily make someone a good singer or musician. A gifted pianist may be a terrible dancer, and a talented ballet dancer won't necessarily make a good flautist. Singing, dancing and playing instruments involve vocal, fine and gross motor competencies, which can all be set at different levels. Our early ancestors' main use for singing, dancing and playing was probably to awaken and express emotions, and to strengthen the cohesion within the tribe – just as national anthems are intended to do today. There are many cultures in which song, music and dance still fulfil this function. In our society, unfortunately, it has mostly been lost. The experience of music has been reduced to performances by individual artists, consumed by millions of people at concerts and over the internet. As of November 2017, the song 'Hello' by Adele has been viewed just over 2 billion times on YouTube. Adele sings of a painful parting and terrible loneliness. It's no coincidence that so many people, who may well have been lonely themselves at some point, feel that this song speaks to them. Music can touch us in a way that no other competency can.

FIGURAL AND SPATIAL COMPETENCIES

Figural and spatial competencies enable us to understand the physical environment, with its outward appearances and spatial dimensions. We use visual perception to recognize the shapes and colours of objects, and the sense organs of our skin, muscles and joints (tactile-kinaesthetic perception) to provide us with information on things like the weight and surface texture of objects. In combination with motor competencies, we use figural and spatial competencies to express ourselves through art. As early as 30,000 years ago, our ancestors were already using these competencies to paint on cave walls, carve ivory figurines and make jewellery (see Chapter 1). Since then, numerous cultures have produced great works of art including paintings, sculptures and architecture. Figural and spatial competencies also have an important place in the sciences. In 1954, James Watson and Francis Crick made an important contribution to unlocking the genome with their three-dimensional model of the DNA double helix. In modern physics, scientists use a multi-dimensional model to reflect the relationship between space and time, and biologists and palaeontologists look at the many different appearances of life forms in order to draw important conclusions about evolutionary processes, such as how one plant or animal species evolved from another (see Chapter 1).

Human children are not born blind like kittens. In the very first weeks of life they begin to recognize certain visual patterns, such as the human face.[47] By the time they're a year old, their figural competencies are so advanced that they can tell the difference between

Figure 5.9: Drawings by Eva (*left*) and Martha (*right*).

familiar and unfamiliar faces and discern a range of emotions from facial expressions (see Chapter 3). They are also able to understand their physical environment down to the smallest detail. At the age of two or three, children start to give artistic expression to their ideas through drawing and painting. At four, their scribbles develop into closed shapes, and they typically draw circles with legs to represent human figures. Over the years that follow, children separate out the human body into head, neck, torso and limbs, and start to give their drawings hair, fingers and toes. In adolescence, their figural and spatial competencies are extended one final time. Perspective is introduced into drawings, and children develop an understanding of representative geometry and maps of the land or the stars. At this point, their figural and spatial development is mostly complete. The drawings by Eva and Martha, both six years old, show that creative ability lags a long way behind figural and spatial perception. Our perception is always much better developed than our creative abilities. The drawings also illustrate the different levels of creative development the two girls have reached. The fact that such differences also exist in adults is illustrated by the results of a study we carried out at Zurich Children's Hospital.

We asked forty medics to copy out the Rey–Osterrieth complex figure as accurately as they could (Fig. 5.10). Fifteen minutes later they drew the figure again from memory. A radiologist achieved the best result; the pattern he drew was very nearly the same as the original (B). The weakest result came from an endocrinologist; his drawing was only as accurate as that of an average ten-year-old child (C).

Spatial competencies tend to be slightly stronger in men than they are in women. As I have already mentioned, anthropologists trace this difference back to the fact that men spent 200,000 years honing their orientation skills as hunters. Women are a little more talented in figural competencies, such as following knitting patterns. But – as with all competencies – the difference between these groups is much less significant than that between individuals. Some skilled female orienteers can find their way through difficult terrain perfectly well using rudimentary maps. And some men are incapable of reading a street map and rely on their satnav when driving.

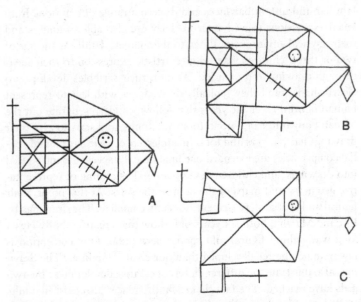

Figure 5.10: The Rey–Osterrieth complex figure. Forty scientists copy out figure A. Fifteen minutes later they reproduce the figure from memory. B: best reproduction; C: weakest reproduction, corresponding to the average performance of a ten-year-old child.

LOGICAL AND MATHEMATICAL COMPETENCIES

All knowledge is provisional until it is proven wrong.

Galileo Galilei (1564–1642)

Most people regard logical and mathematical competencies as the very pinnacle of human intelligence: thinking *par excellence*. This may well be because of all the mental constructs logical and mathematical knowledge has proven the most consistent and free from contradictions. This viewpoint is reinforced by the fact that logical and mathematical competencies form the basis for science and technology, which have provided us with a wealth of incredibly useful knowledge and products.

In the broadest sense, logical and mathematical competencies consist of discoveries about the nature of objects and their interaction. These discoveries are made through repeated, precise observations, the creation of non-contradictory qualitative and quantitative criteria, and a systematic use of these criteria. One reason why we are seduced by logical and mathematical thinking lies in the idea of proof. Observations are always being mistaken for immutable truths. The philosopher Karl Popper – following in the footsteps of Galileo Galilei – relativized the status of scientific knowledge by saying that empirical and theoretical insights could never be completely validated, only completely disproved.[48] And this goes not only for scientific knowledge, but for the observations we make in our everyday lives.

An understanding of numbers (arithmetic) is one of the building blocks of our logical and mathematical competencies. It involves quantifying sizes and amounts, using numbers operatively, and understanding their qualitative features: natural numbers, prime numbers or π (pi), for example.

As an academic discipline, logic and mathematics involves a high level of abstraction. But its roots lie in the concrete experiences that children have with their physical environment as they develop. In the first year of life, for example, a child gains an understanding of cause and consequence. She discovers that when she pulls the cord of the music box, music will come out. At 18–24 months, she realizes that objects can be the same or different depending on certain qualities such as colour or shape. This insight is the start of categorization, another basic element of logical thinking. The simplest ideas of quantities can be observed in newborns and small babies.[49] But, until they reach primary-school age, children don't understand the consistent quality of amounts and volumes, as Jean Piaget has shown.[50] In one of his studies, he filled containers of various different widths, depths and heights with the same amount of liquid and asked children between the ages of three and seven which container had the most liquid in it. They answered unanimously that it was the one in which the surface level was highest.

The children only considered the height of the liquid, and not the other two dimensions that determine volume. Children only understand that the depth, width and height of the containers are equally

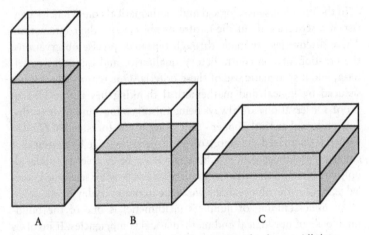

Figure 5.11: Constancy (invariance) of amounts and volumes. All three containers hold the same amount of liquid. Children under the age of seven believe that container A holds the most liquid, because the surface level is highest. (Based on Piaget 1997)

important once they have explored the relationship between the three dimensions of the containers and their contents by pouring the liquid from one to another repeatedly. You can't reduce this kind of learning process to the formula 'volume = width × height × depth'. Piaget's experiment shows that only experience leads to a genuine gain in knowledge. The amount of experience children and adults require before they have really grasped a connection differs from one person to the next.

At the time they start school, most children also lack an understanding of numbers that is sufficiently developed for them to measure physical sizes and look at them in relation to each other. Over their school career, their understanding of numbers is extended gradually, until they have a quantitative grasp of spatial dimensions, amounts and volumes. Here, too, practical experiences such as measuring the classroom are indispensable. Logical thought remains strictly concrete right up to the start of adolescence, meaning that children can only deal with physical objects and the mental images of those objects. Abstract thinking only begins at puberty. Young people then understand logical connections and are able to cope with abstract concepts, such as those involved in algebra.

The logical and mathematical competencies develop differently from one child to another and reach different levels in adults, too. At eight years old, some children have an understanding of numbers that others only achieve at sixteen, or not at all. In adulthood there are thinkers such as Albert Einstein, who came up with the theory of relativity, and people of otherwise average intelligence whose understanding of numbers is so limited that they pay for their shopping with large-denomination notes, because they can't work out the correct change. We can get an idea of how overwhelmed these people must feel when we go on holiday and have to deal with an unfamiliar local currency.

TEMPORAL AND PLANNING COMPETENCIES

> What then is time? If no one asks me, I know: if I wish to explain it to one that asketh, I know not ...
>
> Saint Augustine (354–430)[51]

We are always available, always rushing from one appointment to the next. The day is all planned out – and that goes for children as well as adults. Time pressures are everywhere: in our families, in society and in business. High time, then, to review our temporal and planning competencies.

Human lives are measured by the ticking of the clock. But that hasn't always been the case. For our ancestors, and for plants and animals, the rhythm of life was determined by sunrise and sunset, the differing lengths of day and night, and the seasons. The great change only came with the start of industrialization in the nineteenth century. Since then, industry has increasingly imposed its own diktats on time. We retain certain cultural customs and rituals like Easter, harvest festival and Christmas, which have helped to mark the passing of the year for centuries. But we are mainly ruled by the rhythm of the environment we have created. Artificial light, which turns night into day, the dynamics of the workplace and global twenty-four-hour communications have all contributed to an extreme acceleration of our lifestyle.

Our forefathers started using devices such as sun, water and candle clocks to measure time thousands of years ago. The invention of the mechanized clock in the fourteenth century meant that units of time were no longer determined by nature, as with the sun dial, but measured out by humans. Today, the units of time that modern technology is able to measure go far beyond our subjective experience. High-precision measuring devices such as quartz and atomic clocks can capture time down to the nanosecond, while enormous telescopes can measure huge periods, such as the age of the universe. Without these modern devices, technological progress – information technology, for example – would have been unthinkable. The measurement of time also plays a crucial role in disciplines such as anthropology and life sciences, which study developmental processes. And in the economy time has become just as important as money.

But time is also measured by powerful internal clocks which are much older than the external clocks we have created – primeval in the truest sense.[52] From single-celled organisms to the most highly developed animals and humans, all life is ruled by biological rhythms. They regulate biochemical processes on a cellular level and physiological processes like breathing and heartbeats. At the centre of our brains, in the hypothalamus, we have an inner clock. It gives us the circadian rhythms that regulate bodily functions, including the sleeping and waking cycle and the release of hormones. These circadian rhythms last an average of twenty-four hours – the term is taken from the Latin *circa* (about) and *dies* (day). But the real rhythms of our lives are now placed under so much stress by external clocks that it is increasingly affecting our physical and mental wellbeing.

Humans are probably the only living things with a conscious – though also very subjective – sense of time. We experience time differently depending on the situation in which we find ourselves. Periods of time that are filled with activities (especially if these activities are enjoyable) seem short, while periods where little happens feel torturously long. In our memory, the exact opposite is true: we feel as though eventful periods were long, and uneventful ones were short.

Our ideas of time are closely tied to our consciousness. We cannot banish time from our consciousness, and nor can we manage our time without conscious thought. We experience time as a state of flux. The moment we are experiencing right now is the present, while we describe what lies behind us as the past and what lies ahead as the future. With these descriptions, we give time a direction and a kind of spatial dimension. But the grid of time is to be found less in the conscious than the unconscious mind (see Chapter 6). It is laid over our experiences, memories and activities in the unconscious, and, when they enter our consciousness, this grid gives them a place on a subjective axis of time.

Every age of man comes with its own sense of time. Newborns and babies live by their bodily rhythms, their breathing, heartbeat and digestion. After a few weeks, when their sleeping and waking pattern has adjusted to the rhythm of day and night, their waking periods are given a set chronological order, involving activities like eating and playing. Babies and toddlers live a mainly 'timeless' existence in the here and now. At the age of three or four they start to become conscious of time, and this consciousness is gradually extended over the years that follow: before and after a meal; before and after a nap; morning and evening; yesterday–today–tomorrow. By the time they are five, children have developed a limited understanding of past, present and future, which enables them to understand and use the various tenses of verbs. At six or seven they learn the days of the week and the seasons, and then they start to think about weeks and months. Between the ages of seven and ten they learn to tell the time. With its complex division into hours, minutes and seconds, this is a real challenge for children, and presupposes a well-developed understanding of numbers. By the time they enter puberty, children are capable of taking an overview of life from birth to death and understanding large periods of time such as past and future centuries and millennia – in history lessons, for instance.

Very little research has been done on whether everyone has the same sense and understanding of time. But we can assume that it differs from one person to the next, because our ideas of time are formed in combination with other competencies, such as our understanding of numbers. A sense of time is also shaped by the environment in which an individual lives, their previous experiences, and their expectations of the future. Differences in people's temporal competencies are also revealed in their planning skills. The better someone is at taking an overview of a chronological sequence of events and estimating the time an activity will take, the better they will be at organizing their everyday lives. Everyone uses and plans time in their own way, and finds planning activities more or less stressful.

Our ancestors developed ideas of time that fitted their living conditions and the prevailing cultural and social realities of their age. Today, between the permanent flood of communication and the globalized economy, we have much less control over our ideas of time and how we use it. It has become difficult to strike a balance between the many and varied demands on our time that we face every day, our individual circadian rhythms, and our need for rest.

MOTOR AND KINAESTHETIC COMPETENCIES

Eagles fly, reindeer run and fish swim. Humans play football, climb rock faces, ride bicycles and jump on trampolines. And this by no means exhausts our repertoire of motor activities. We make models, paint and draw, play the violin or a percussion instrument, and dance. Over the course of human history, the range of activities for which we use our motor skills has become extraordinarily diverse.

The reason our motor and kinaesthetic competencies are so varied is that they are used in combination with other competencies, depending on what we want to achieve. Good footballers don't just need to manoeuvre a ball well; they also need spatial awareness and tactical playing skills. A violinist doesn't just need dextrous hands; she has to have excellent musical competencies as well. The combination of motor abilities with other highly developed competencies is just as important for specialists like surgeons or graphic designers.

And we don't just use our motor and kinaesthetic competencies to express ourselves and have an effect on our environment. They also aid our tactile and kinaesthetic perception (sensitivity to surfaces and depths). When we reach for an apple, the sense receptors in our skin and muscles tell us how large, hard and heavy the fruit is. When we walk around a house, our kinaesthesia (sense of movement) and visual perception give us precise information about the direction in which we are moving and how a room or a staircase is configured. All this information comes from receptors in the muscles and joints. And our sense of position and balance (provided by the vestibular system) also gives us information on the position of our body in space; for example, whether we are standing up or lying down.

In the first years of life, motor and kinaesthetic development mainly involves acquiring abilities such as attitude control, locomotion and grasping.[53] In the years that follow, children acquire skills that might include riding a bike and writing. The development of motor abilities is completed during adolescence. Adults can still learn skills like playing tennis, of course – though the learning process is more difficult and less thorough than it is for children (see Chapter 3).

Motor abilities and skills are developed to different levels even in early childhood. Some children take their first steps at ten months, while others only manage it at twice that age. Some seven-year-olds achieve things in the area of motor skills that other children can't do before the age of sixteen (see Chapter 3). Levels of ability in activities that use fine motor skills, such as painting and drawing, or gross motor skills, such as football and gymnastics, vary wildly between individuals of all ages. The broad range of motor achievements of which humans are capable comes not just from our individual levels of motor and kinaesthetic competencies, but from the complex way in which they are combined with other competencies.

PHYSICAL COMPETENCIES

On the neo-natal ward, Sara is the favourite with all the nurses. She's the one who gets picked up and carried around the most. The way Sara snuggles into a nurse's chest is so adorable. She has a friendly, open face and a funny shock of hair, and she smells so good.

Childcare professionals talk about children with different degrees of emotion. They will be embarrassed if you point out that they have favourites, because they try to treat all children equally. But most people feel this way – parents sometimes even have favourites among their own children. Once they're old enough to start school, children will have different levels of appeal for their teachers; a fact reflected in the attention they receive and even in their grades.[54] In adolescence, girls with an attractive body shape and a beautiful face will be the ones who turn the boys' heads. And, as adults, physical appearance plays a role in much more than just finding a partner (see Chapter 1): on the whole, tall, good-looking people do better socially and professionally. And older people can still have a strong physical charisma, which influences their families and carers.

Physical attributes can influence us in our relationship behaviour to such a degree that we allow ourselves to be guided by prejudices. Schoolchildren and even some adults regard unattractive people as less intelligent.[55] We also make a connection between beauty and morality: a good-hearted princess has to be pretty; an ugly witch has to be evil.[56]

Whether we are prepared to acknowledge it or not, physical appearances and clothes say a lot about a person, and influence our attitude towards them. It makes a difference if the person we're talking to is a head taller than us or we are looking down on them. The way someone is dressed can make us feel admiration, suspicion or even aversion. We deliberately use our bodies and how we dress them to gain social attention in situations like job interviews and dates. And we associate our physical appearance with ideas such as beauty and manliness.

At every age, our appearance is made up of numerous elements, including facial features and eye colour, hair colour and style, height and physical proportions. People make a huge effort to increase their physical attractiveness through grooming, cosmetics and fitness. The way someone dresses also tells us how they want to be perceived, and often says something about their social status. Even small children realize that they attract attention in pretty skirts and cute dungarees. In adolescence, clothes become a status symbol. The way young people dress and use accessories signals that they belong to a particular clique.

Scent is also important, though we're rarely conscious of it. Perfumes have been widely used in nature for millions of years. Some plants use seductive scents to attract insects that will spread their pollen or seed; others ward off creatures that might eat them with a foul stench. Animals use scents to interact with others of their species and arouse the interest of sexual partners. Dogs mark their territory by urinating in specific places, and butterflies can attract potential partners from several kilometres away. Humans also have scents – and not just the pungent smell of our sweat. Pheromones, which we only perceive unconsciously, play an important role in interpersonal behaviour. Everyone has their own body scent, which is genetically determined and of far greater significance for our relationship behaviour than we might think. The huge number of perfumes and deodorants that humans use, depending on the culture in which they live, tells us how just important smells are to us. Worldwide, the annual sales of scent are estimated at $20 billion.

Taste is also part of our physical competencies. The taste buds on our tongues can detect five qualities: sweet, sour, salty, bitter and umami (a 'meaty' flavour, brought out by glutamate). For animals, the biological significance of taste is less to do with communication within or between species than with finding food and assessing how edible it is. Humans have developed their sense of taste and smell to do much more than this. Talented chefs use their skill and imagination to turn food and drink into a gourmet experience. They select prime fruit and vegetables, and enhance dishes with an array of different spices. Here, too, a range of competencies come into play: the sense of smell and taste; an instinct for choosing and combining foods; figural and spatial abilities for arranging it on a plate; and social skills for presenting it to diners.

Eroticism and sexuality form another important physical competency. In adolescence, our physical appearance acquires its lasting social significance as we develop secondary sexual characteristics and gender-specific body proportions. Sexual charisma, the desire to be seen as sexual beings and the urge to satisfy our sex drives remain important to us right into old age.

Physical competencies are much more important to some people than they are to others, and we all use and experience them in highly individual ways. Some people place great value on how they look, while others neglect their appearance almost entirely. They receive their recognition in other ways – through their social competencies,

Figure 5.12: Anna plays 'family dinner'.

for instance. The effort we make in the area of eroticism and sexuality also differs greatly between people. Some work hard on their sexual charisma and attractiveness, while others pay no attention to it at all.

In general, physical competencies are not accorded a particularly high value. They are decried as sensual or even morally dubious, and dismissed as mere 'outward appearances'. But they play an important role in our day-to-day lives, from birth through to old age. Even so, we should take care not to overestimate physical competencies. The way that people use social competencies such as facial expressions, language and motor skills also contributes to how attractive they are. And the impact of physical attractiveness decreases the better we know someone, while other competencies become increasingly important.

THE UNIQUE INTERPLAY
OF COMPETENCIES

Anna is playing 'family dinner'. She arranges the chairs around the table and sits the dolls in the places that her mother, father, sister and she herself usually take for dinner. She chooses the right number of toy plates and cups, and little knives and spoons, and distributes them using well-honed fine motor skills. She plays the roles of parents and children with the dolls, mimicking how they speak. She serves up pretend food, makes the dolls eat and drink, and then clears the table. At the age of twenty-six months, Anna is achieving incredible things. Her game doesn't just use five competency areas – figural and spatial, motor and kinaesthetic, social, linguistic and planning – it also brings them together.

As children develop their individual competencies, they also weave them together. The huge networks formed during this process may be the most plausible explanation for why the human brain is so extraordinarily large in comparison to our closest relatives, the primates, and why human brain development takes fifteen years (see Chapters 1 and 3).

In their activities, adults frequently use several competencies at once. A teacher doesn't just transmit subject knowledge. He also uses his social competencies to motivate his students, and his linguistic competencies to communicate knowledge and pass on skills. He uses his figural and spatial competencies to create graphics that make the relationships between things easier for students to understand. And he uses his temporal and planning competencies to structure individual lessons, teaching units and the plan for the school year.

The way children and adults live their life is dependent on how their profile of competencies is configured. Some children play games that use their social and linguistic competencies, while others occupy themselves more with the spatial arrangement of objects. Adults also lead different lives depending on their individual profile of competencies. The following illustration shows the competency profiles of two women, both of whom are around forty years old. Sofia is married to a city mayor and has two children. Erika is a graphic designer and artist. She is single and leads a modest, secluded life.

Sofia puts her well-developed social and linguistic abilities to use in caring for disabled and socially disadvantaged people. Erika's linguistic competencies are average, but her social competencies are low. She dreads having to make small talk at exhibition openings. She is a very keen dancer. Sofia sings in a church choir. She draws very clumsily and her spatial awareness is so limited that when driving she has trouble finding her way without her satnav – even in the city where she lives. Erika, by contrast, has exceptional figural and spatial

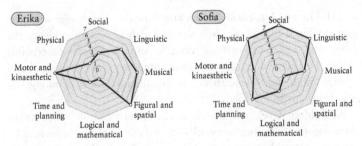

Figure 5.13: Individual competency profiles for Erika and Sofia.
Competency ratings: 1 = very low; 4 = average; 7 = very high.

perception. She is in her element with any creative activity, whether it's painting, making pots or sculpting. Like Sofia, Erika isn't very competent in maths and is always getting into financial difficulties. Her method of working is chaotic. Preparing work for an exhibition regularly places her under extreme stress. Sofia, on the other hand, loves planning and organizing events for her husband, the mayor. Her motor competencies are not very well developed, while Erika's fine and gross motor skills are excellent. Erika is slim and her looks are unremarkable. Sofia is attractive and always very well groomed.

The level to which people's competencies are developed depends on their genotype and the experiences they have had, particularly in childhood (see Chapter 3). The reason Sofia speaks four languages is not just because she has very good linguistic competencies. Her parents worked as overseas missionaries, and she grew up in several different linguistic cultures. Erika was able to develop her figural and spatial competencies fully as a child because her parents were impressed by her creative zeal and made every effort to support her endeavours. Erika and Sofia are lucky enough to be able to live lives that fit their competencies.

As we saw in Chapter 3, children are selective in seeking out the experiences they need to develop their own individual competencies. Adults behave in exactly the same way. In their work and private life, they look – often unconsciously, as children do – for places where they can make the best use of their competencies. The differences between Erika and Sofia's competency profiles show that an individual can't lead just any life. If Erika and Sofia each had to inhabit the other's role, they would both be under- and over-challenged in different areas. From the Fit Principle point of view, mental and physical wellbeing, a sense of self-worth and self-efficacy all depend on the extent to which a person succeeds in achieving harmony between their competencies and their environment – just as we saw with the satisfaction of needs.

FUNDAMENTALS FOR
THE FIT PRINCIPLE

We all have a unique jigsaw puzzle of competencies

In September 1896, the eighteen-year-old Albert Einstein took his higher school-leaving exams at the Aargau Kantonsschule in Switzerland. In algebra, geometry, descriptive geometry and physics he gained the top grade: 6. In geography, artistic and technical drawing he got a 4, and in French he only managed a 3 ('poor').

People often say that Albert Einstein, perhaps the most famous physicist of all time, wasn't particularly good at maths and physics when he was at school; in fact, they say, his grades were terrible. But this claim is based on a misunderstanding: Einstein went to school in Switzerland, where the top grade is 6, not 1 as it is in Germany. Albert Einstein's exam results range from 6s to 5s and 4s, and even a 3. Einstein was extraordinarily gifted, but he was no polymath. People whose competencies are developed to an above-average standard in several areas are very rare. Leonardo da Vinci was one of these exceptions. He achieved extraordinary things in painting and drawing, architecture and engineering, poetry and the natural sciences.

We all have our own unique jigsaw puzzle of abilities and skills. A two-year-old child can be very advanced for his age socially and linguistically, while his motor abilities are very underdeveloped. Another child of the same age can be the exact opposite. Some adults have tremendous social abilities but very poor logical and mathematical abilities. Others are highly logical and skilled in maths, but less good socially. People who have a disability in one area and are very gifted in another show us just how far apart our competencies can lie. In 2004, Daniel

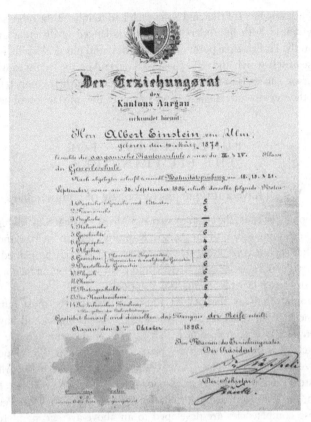

Figure 5.14: Albert Einstein's matriculation certificate.

Tammet made headlines with his mathematical abilities when he managed to recite π to 22,514 decimal places from memory.[57] He has an astounding mathematical talent, but at the same time his social competencies are very limited. Neuropsychologists have diagnosed him as a high-functioning autistic savant.

Just as some people have unusually high levels of ability, there are also those who are less able. A 'specific learning difficulty' can be diagnosed when a person's reading competence or understanding of numbers is much less developed than all their other abilities. These learning difficulties can affect any competency, and occur more frequently than is generally thought. At least 30 per cent of the general population is estimated to be affected by a specific learning difficulty. How severely these people are affected by their low ability depends on the challenges that society and the world of work throw at them. Difficulties with reading or maths are much less of a disadvantage in a farming community than they are in a modern service society.

Part of the reason our talents are so diverse is that even the separate elements of a competency can be developed to different levels within an individual. Let's take social competencies, for example. An actor's nonverbal communication and social learning skills are extremely well developed, giving her a very expressive face and enabling her to reproduce various human character traits convincingly. A nurse has a high level of empathy and caring behaviour, which suits her to the work of caring for children, the disabled and older people. A philosopher's strengths, meanwhile, lie in the area of social cognition. She is an expert when it comes to ethical questions in medicine and business. She would be much less well suited to being an actress or a nurse. The actress and the nurse aren't suited to a career in philosophy, either.

As with our basic needs, the unique make-up of our competencies is part of our individuality. We should be grateful for those competencies that are well developed, and accept those that are weaker. And, as with the basic needs, our task is not only to accept our own profile of competencies, but also to respect other people's. Our children, our partners and every one of our fellow humans have their own profile of competencies, which we should bear in mind when thinking about what we expect or require from them.

A world where we can develop and use our competencies

We are not polymaths. Our competencies developed over hundreds of thousands of years during which humans lived in small communities, surrounded by nature. Now we have largely alienated ourselves from the natural world and live in anonymous mass societies, where we are less and less able to develop and use many of our competencies – sometimes with serious repercussions.

Let us turn first to children. Children want to develop all their competencies, not just those that their parents and teachers demand of them. For that, they need to grow up in an environment where they can have experiences suited to their level of development. And children don't just want to develop their competencies: they want to be allowed to learn from experience in their own way and at their own pace. In today's education system, where children's learning is largely directed by other people, this is hardly possible. How children learn is just as important as what they learn.

At pre-school age, children spend a lot of time in a bedroom filled with toys. They sit in front of the television – in Switzerland, children between the ages of two and four watch an average of 1–2 hours per day – and when they're older they play on smartphones and tablets. They are denied many important experiences, much to their disadvantage. Their motor development will be hugely affected by whether they can run through fields and climb trees every day or just walk from room to room, where their greatest motor challenge is climbing on to a chair and not tripping over door sills. Children would much rather play by streams and lakes than be penned into play areas where their safety is guaranteed. And it comes as no surprise that an increasing number of children are being diagnosed as hyperactive, when they have too little opportunity to run around outside. Children are also unable to develop their social competencies fully, because they often lack sufficient interpersonal experiences. Some parents have so much stress in their own lives that they don't spend enough time with their children, playing or going for walks. Children also increasingly lack stable and trusting relationships with neighbours and relatives who might serve as role models. But above all they are missing out on time spent with other children. Children's social development relies on them spending a lot of time – several hours a day – with other children of various ages, and learning from them. Children who don't go to a playgroup or nursery, and have no other children to play with, miss out on many of these experiences, and their social and linguistic development is badly affected as a result.

And then children go to school. A child-centred education system would allow children to develop all their competencies independently. Instead, children suffer from a general obsession with hothousing. They are supposed to acquire the skills that are needed in society and industry. Swiss industry is currently in need of more scientists and IT specialists, and so schools eagerly increase the number of hours devoted to science teaching, and introduce programming to the curriculum.

Education in schools is largely limited to teaching students skills like reading, writing and using computers, and providing them with the knowledge that education ministers think will be useful to them later in life. The focus is on attainment rather than developing competencies. From the first day of primary school to the time students graduate from universities and apprenticeship programmes, a detailed curriculum dictates what they have to learn. In primary schools, the focus is on rote learning, doing homework and passing exams. And in colleges and universities, students still spend most of their time learning huge amounts of material by heart, collecting credits and certificates and doing their best to graduate *summa cum laude*.

Is the education system successful? Businesses are stoking the obsession with hothousing children, but they are also complaining that young adults don't think independently, and wait to be told what to do. They lack any kind of initiative, we are told, and shy away from responsibility. And, even more worryingly, many school and university students feel their learning isn't under their control; they are overstretched and stressed. Their physical and mental wellbeing is being impaired and they are suffering from various complaints, including headaches, sleep disorders and exhaustion.

And what has become of civil society's educational canon? Secondary school students find it difficult to warm to Goethe's *Faust*, despite their teachers' best efforts. Philosophers like Socrates and Kant are met with nothing more than a weary smile. Students know the categorical imperative, but they doubt that reading Kant's *Critique of Pure Reason* will offer insights that have any real relevance to their lives. Young people certainly have values, and they feel them strongly. But these values don't come from the nineteenth century any more. Children are concerned about justice in a world ruled by poverty, hunger and misery. Why is there so much injustice, and how can they make the world a better place? Students want their teachers to discuss these things with them as equals. And, above all, they want to know what the real meaning of life is.

From the Fit Principle point of view, the education system needs rebuilding from the ground up. It can no longer exclusively serve the needs of industry. It must focus on children's individual development, and their basic needs and competencies. At the end of their education, every young adult should be able to say: I have been able to develop all my competencies, and have achieved most of the things I set out to do. I have a good sense of self-worth and good self-efficacy, and I am looking forward to my future life as a member of society and a contributor to the economy.[58]

Adults, after all, don't want to work just to earn a living. They also want to use their competencies and achieve meaningful, satisfying things. In today's society and economy, this is becoming increasingly difficult. Competencies that were once almost essential have largely lost their significance. Our ancestors used them for practical activities such as hunting and gathering plants, or building huts and producing tools and everyday objects like baskets and clay pots. Meanwhile other competencies, such as the logical and mathematical competencies used in information technology, have become very important. And even highly developed competencies aren't always enough in today's world. They are suited to a life in small communities made up of familiar people, and not to an anonymous mass society. Until 150 years ago, most people worked as farmers and craftsmen. In Europe and the USA today, just 2 per cent of the population is employed in industrialized agriculture. Artisanal occupations are also in decline,[59] while 70 per cent of people work in the service sector, where social and linguistic competencies are paramount. But what has happened to all the people who are good with their hands and want to do physical work? Can a talented carpenter or a powerful labourer become a capable IT specialist or a social worker, just like that? This one-sided development increasingly means that, in order to earn a living, some people have to perform tasks that are very difficult for someone with their set of competencies. They suffer from tremendous stress at work, and even fear they won't be able to find a job. A nightmare scenario? Only if we don't take a serious look at the current situation and rise to the challenge it poses: we need to reform society and the economy so that people can satisfy their basic needs and develop and use their competencies (see Chapter 10).

6

Our Ideas and Beliefs

*Humans are the only living creatures that have to explain
the world to themselves in order to live their lives.*

He who has a Why for living can bear almost any How.
Friedrich Nietzsche (1844–1900)[1]

Humans are far from the only living creatures to have ideas about
their environment. All the more highly developed animals have the
ability to visualize things. A cat knows exactly what the tom that
wants to take over its territory looks like. A squirrel has a good spatial
idea of the area in which it lives, which means that in winter it can
find the stores of food it hid in the autumn. Animals' ideas are sophisticated, but selective. They only ever reflect one part of the environment,
namely the part that is important for the animal: a feeding ground or
a creeping tomcat.

We humans see much more than just the things that will help us
satisfy our elementary needs. We want to gain as complete an understanding of the environment as possible, and we develop broad ideas
to help us in this. When we're confronted with a puzzle, we can't rest
until we've solved it. Occasionally we even claim to have discovered
the ultimate truth. We are probably the only living creatures that
have to explain the world to themselves in order to live their lives. Or
as Nietzsche said so astutely: having a Why for living helps us to bear
the vicissitudes of life.

There is another peculiar aspect to our ideas. We can make ourselves
conscious of some (but not all) of our ideas. We can think about them
and communicate them with the help of language. The ability to have

ideas about ourselves and the world, and the need to share them with other people, was probably one of the principal reasons why humans developed a highly sophisticated form of communication and some very successful survival strategies along with it.

In the following section we will try to clarify what we mean by ideas and how they are formed. Why are our ideas so ambiguous? And why do we become conscious of some ideas, while others remain buried in the unconscious? What is consciousness? Why does every human being have their own ideas, and why are collective ideas like religion or ideology so powerful?

THE ESSENCE OF OUR IDEAS

Ideas through the ages

Humans have come up with their own ideas about the world in every culture and every age. Our ancestors, who until 70,000 years ago lived in central Africa, probably thought the world was made up of savannahs, jungles, mountains, valleys, rivers and lakes. Their descendants, who migrated across Africa all the way to Eurasia, discovered that wasn't all: the world contained a lot of salty water, too. They adapted their ideas accordingly: the earth was a flat slice of land surrounded by a sea that stretched to the edges of the world.

As early as the fourth century BC, scholars came to the conclusion that the earth was a sphere. The Greek philosopher Aristotle came up with three well-founded arguments for this view. First: when a ship sails out to sea, the hull vanishes first, and then the sail. Second: during a lunar eclipse, the shadow of the earth that appears on the moon is always round. And third: in southern lands, southern star constellations are further above the horizon than they are in the north. All the same, the vast majority of humans stuck with an idea that was a much better fit for the reality they experienced: the earth is a flat disc.

The explorer Christopher Columbus also believed the earth was a globe, and was thus convinced that if he sailed west from Spain, he would reach China and India. In 1492 he took three ships and set sail towards the setting sun. But, as we know, he didn't reach China: he discovered America. His fundamental error was his failure to imagine that another huge continent might lie between Europe and Asia. In 1519 the Portuguese explorer Ferdinand Magellan set off with five ships to circumnavigate the globe. The voyage cost him his life, but another member of his expedition, Juan Sebastián Elcano, completed the first circumnavigation in 1522. This proved that the world was a sphere once and for all, and people adapted their view accordingly.

But then the accepted ideas about the earth were called into question once again. This time the issue was the planet's position in the cosmos. At the beginning of the seventeenth century Nicolaus Copernicus, Galileo Galilei and Johannes Kepler came to the unanimous conclusion that the sun didn't orbit the earth: the earth orbited the sun. The heliocentric view came as a shock to humanity – but the earth was set to be marginalized still further. We now know that our solar system isn't the centre of the universe, either: it lies on the edge of the Milky Way. And the Milky Way is just one of at least a hundred billion galaxies, each of which contains around a hundred billion stars, circled by billions of planets.

What can we learn from this short history of humanity's view of the world? In all cultures and all ages, humans have developed ideas about the world that made sense to them, based on what they saw and experienced. When new insights were gained, they adapted their image of the world. The process of acquiring knowledge has been very similar in the sciences, including biology, physics and psychology. Our image of mankind – like our image of the world – has always been an expression of a particular culture and time. And we are just the same as people throughout human history. We are always adapting our ideas to fit the changing circumstances of our lives. The small insights we gain in our day-to-day lives are always only provisional, like those of the astronomers. But that doesn't stop us thinking of the current state of knowledge as the final word on a subject: we can't imagine what the future might hold.

How children arrive at their ideas

I observed children at play over more than three decades, and gradually learned to understand how they arrive at their ideas. Let us take
the example of how spatial ideas develop from birth to adulthood.

In the first weeks of life, a baby has a very limited perception of
space. He can only focus on things up to 20 cm away. By about four
months old, he can focus on things at different distances. He starts to
reach for objects, watches his parents and siblings when they walk
around the house, and learns to judge distances. At the age of 6–12
months, he starts to crawl and actively explore the space around
him. Up to this point, the development of his spatial perception has
been almost exactly the same as it is for the young of the more highly
developed mammals.

But then he starts to develop in a way that is characteristic of humans:
his games become an active engagement with the spatial relationships
between objects. At 12–18 months he starts to take an interest in
containers and what they contain. He will put wooden blocks into
a box and take them out again. At the age of eighteen months he
becomes fascinated by verticals and builds a tower out of the blocks.
Six months later, his interest moves on to the horizontal plane. He
now places the blocks in a row and enjoys playing with his train set.
At around three years old he combines the horizontal and the vertical
and builds steps. At three or four the child brings all three dimensions into his game. He might construct a Lego garage for his toy
cars.[2] At the age of two, children are already capable of linking their
spatial ideas to the corresponding linguistic terms. First, they apply
the preposition 'in' to the idea of container and contents. Then they
develop an understanding of the prepositions 'on', 'under', 'behind'
and 'in front of'.[3] Over the years that follow they put their creative
abilities to use by making copies of their physical environment; for
example, using chairs and sheets to construct a little house. The development of spatial ideas is completed in adolescence, when children
become capable of abstract thought. In geometry, students move from
two-dimensional to three-dimensional representation. They learn how
to work out the circumference of a circle or the volume of a sphere
by measuring its diameter. This creates the cognitive conditions for

understanding the laws of space and making practical use of them. In physics, students learn the rules of optics by experimenting with lenses. In the process, they develop an understanding of microscopes and telescopes, which in turn gives them an insight into the structures of cells and the dimensions of outer space. Using four competencies, the table shows the ideas and linguistic terms that are formed as children develop, and the activities and products that result from them.

Competencies	Ideas and linguistic terms	Activities and products
Social	Empathy, love, sympathy, social rules, morality	Rituals such as weddings and baptisms, caring, legal systems
Linguistic	Grammar, syntax, vocabulary	Writing, telephones, internet, media
Figural and spatial	Shapes, colours, dimensions	Architecture, sculpture, astronomy
Logical and mathematical	Categorization, causal thinking, understanding numbers	Information technology, sciences, mathematics

How experiences shape our ideas

Ideas, which we might also describe as mental images or schemata, are formed in childhood out of our experiences with our physical and social environments. Children may be able to learn abstract statements such as mathematical formulae, but they will only really understand them once they've had the necessary experiences. Students don't truly grasp Pythagoras' theorem until they've drawn right-angled triangles of various sizes, cut out the squares on the hypotenuse and the short sides, and looked at their areas in relation to each other.

The extent to which experiences shape ideas is particularly clear in social development. 'What is pity but the act of feeling another person's misery in our hearts, which compels us to give what help we can?'

(Saint Augustine, *De Civitate Dei*, IX, 5.) Simply telling children that compassion is a virtue won't turn them into empathetic creatures. You have to give them a chance to experience the feelings and behaviours that lie behind words such as altruism, sympathy or pity, and have an emotional response to them, before they will understand and act accordingly. A child internalizes her parents' caring behaviour towards her, and towards other people; for example, the elderly or disabled. The way her parents talk about politics at the dinner table – their attitude to the plight of refugees, for example – will be internalized, too, but to a much lesser extent than their caring behaviour. Parents and carers are role models, for both good and ill, who shape a child's social behaviour and values (see Chapter 5, 'Social learning'). And the same goes for the way children are taught in school. A teacher dictates the following simple definition of justice for her students to write down: 'Justice is done in a community when there is an appropriate, impartial and enforceable balance of interests, and a balanced distribution of opportunities and goods.'[4] Once students have dutifully learned it, they can hope for good exam results. But they will only act in a fair and just way if they are also treated fairly and have internalized upright and honest behaviour from their role models.

Experience determines the attitudes and behaviours of collectives as well as individuals. In Switzerland, referendums have repeatedly shown that a hostile attitude towards foreigners is more prevalent in areas where people have less direct experience of them. The inhabitants of rural areas, whose knowledge about foreigners is entirely based on hearsay, are significantly more hostile than those in urban areas, who have lived alongside foreigners for many years. A lack of experience gives rise to delusional ideas, vague anxieties and a greater willingness to exclude people, while specific encounters go hand in hand with increased solidarity and willingness to help. The same thing happens to refugees and immigrants who arrive in Europe with ideas that are often just wishful thinking. Only when they have some (hopefully positive) experience of a country do their ideas correspond to the reality there.

We can understand the behaviour and ideas of individuals and collectives better if we assume that their ideas aren't just isolated constructs. Rather, they are based on certain experiences and are closely connected to people's basic needs and current living conditions (see Chapters 8, 9 and 10). Someone doesn't become a Conservative or Green politician just because they are convinced by the party manifesto; when they choose a party, it will be because its goals are the best fit for their individual basic needs and competencies, as well as their previous experiences and the situation they are currently living in. Voters also lean towards the political ideas they believe will satisfy their basic needs the best.

Why our inner images are ambiguous

We become aware of just how large and rich the universe of ideas is when we look at a library, or even just flick through an encyclopedia – or if we attempt to imagine the mountain of digital information that currently exists. One feature of these ideas is that they all come with linguistic terms attached. But there are also ideas that tend to evade language and can only be communicated in a very limited way. We find it difficult to describe the heady scent of a perfume or the warmth in the voice of someone we love. Memories often can't be captured with a single term; we have to paraphrase, perhaps using narrative or film techniques. The same thing goes for our dreams. It's difficult to find the right words for vague anxieties and physical malaise. We often struggle to find a way of expressing something, even though we have a perfectly clear image of it in our minds. Just because we don't have the right words, it doesn't follow that the idea doesn't exist.

For many reasons, ideas that can be captured in language are often only superficially clear-cut. A word such as 'cup' or 'apple' is generally understood as it is meant. But some terms, such as 'upbringing' or 'family', have a different meaning for everyone, because they're based on different experiences. Ultimately, the meaning of many ideas is only gleaned from their relationships with other ideas. Someone's 'life story' consists of all the things that have happened to them between birth and death.

'Religion' and 'ideology' are terms that are quite obviously ambiguous and open to different interpretations. Religion brings together an incredibly diverse range of ideas, each of which has its own meaning, as the following breakdown shows:

The afterlife and the supernatural. All religions are born out of the fear of death. The supernatural entered the human imagination when we developed an idea of time and became conscious of our own mortality. Humans grieve for people they have loved, realize that they too will die, and wonder whether and how they will continue to exist after death.

Banishing threat and easing suffering. When humans began to think about time and became conscious of their own existence, they also started to wonder about the causes of happiness and unhappiness. Why are we threatened by natural disasters such as volcanic eruptions? Why does the rain stay away and allow the harvest to wither? Why do people and animals go hungry? Why do epidemics of cholera and plague break out, bringing illness and death, and why must innocent children endure such terrible suffering? Who is responsible for all this evil? We no longer attribute threats like drought and hurricanes to supernatural forces; they are the forces of nature and the results of the damage we have done to the environment. Most illnesses, such as bacterial infections, are no longer seen as a curse placed on us by the gods or our neighbours; they can be diagnosed and treated effectively. But when we reach the limits of modern medicine, many sick people still turn to God or to spiritual and natural healers.

Creation: why we are here. As they developed the idea of time and became conscious of their own existence and of everything else that existed on earth, our ancestors eventually also started to wonder how all these things came into being and who created them. Depending on their culture and living conditions, they came up with a wide range of explanations, and invented grand myths like the creation story in the Old Testament.

Community, solidarity. People rely on a community for their mental wellbeing. Religions can give the faithful a strong sense of community. Coming together in impressive buildings like churches and temples reinforces this feeling. Communal prayer, singing and dancing create emotional and social bonds between followers of a religion. Rituals are used to celebrate or commiserate the important events in people's lives: happy occasions such as weddings or the birth of a child, and more serious events such as the survival of an illness or the death of a community member. Communal experiences promote a strong sense of belonging. One of religion's greatest strengths is that its communities provide a refuge of emotional security and support for their members.

Morality. We cannot overestimate the influence that religions have exerted on inter-personal behaviour over the course of human history. Religions were and still are so powerful in some cultures because their influence on social behaviour and moral values is very closely tied to ideas of death and the afterlife, and supernatural powers. Offences against the moral code ordained by a religion are branded 'sins' and they come with terrible repercussions. However, the major religions also have a morality of compassion and mercy. They tell us we should help the weak and disadvantaged, forgive those who have strayed and accept them back into the community. A philosophy of total benevolence, as practised by Mother Theresa and Albert Schweitzer, goes beyond someone's own community and even their own wider society. Benevolence is an idea that has freed itself from the constraints of the tribe or small community in which caring behaviour originated. It is intended to benefit everyone in the world.

The term 'religion', then, encompasses a whole series of ideas rather than just one. People experience and understand religion differently depending on the significance that their culture attributes to each of its constituent ideas. A similar level of ambiguity exists for countless other ideas, in the humanities and sciences, for example, which can only be understood from their historical origin and the context in which they are used.

CONSCIOUS AND UNCONSCIOUS IDEAS

And when the woman saw that the tree was good for food, and that it was pleasant to the eyes, and a tree to be desired to make one wise, she took of the fruit thereof, and did eat, and gave also unto her husband with her; and he did eat. And the eyes of them both were opened, and they knew that they were naked; and they sewed fig leaves together, and made themselves aprons.

Genesis 3, 6–7

Humans have probably been exploring self-perception, the conscious idea of themselves as independent people, and consciousness as a mental state, for much longer than written sources like the Bible's wonderful description of Adam and Eve's lost innocence suggest. The philosopher René Descartes laid down a milestone in human self-perception in the seventeenth century with the introduction of the term *conscientia*, consciousness, and the famous *cogito ergo sum*: I think, therefore I am. He goes on: 'Since I am still that same I who doubts, I can no longer doubt this I, even if it is dreaming or fantasizing.'[5] Ever since the Enlightenment, philosophers, psychologists and more recently neuroscientists have been trying to clarify the term 'consciousness' – an effort that has been dogged by many differences of opinion. We still find the idea of consciousness difficult, but we have even more trouble with the unconscious mind. We also overestimate our conscious ideas, and underestimate those that influence our lives from the regions of the unconscious.

The enigma of consciousness

At some point in their lives, almost everyone wonders: what is consciousness? In the past, the majority of people understood consciousness as a mental state reaching beyond the physical boundaries of the individual – and some people still think of it in this way today. We might think of consciousness as a state of being 'animated' or 'ensouled'. Dualists separate the mind or soul from the body. Near-death experiences are interpreted as proof of an extended, transcendental consciousness. The existence of a supernatural plane beyond the physical world is also an article of faith for most religions.

Clinical medicine, by contrast, takes a very pragmatic approach to consciousness. Simple behavioural criteria are used to differentiate between states ranging from full consciousness to coma. When a patient is responsive and knows where he is in space and time, he is awake. If he's disoriented and incapable of conversation, but still reacts to stimuli like touch, his consciousness is restricted. If he doesn't even respond to strong pain stimuli, he is comatose. Neurophysiology divides consciousness into various levels of sleep and wakefulness, which can be measured with the help of an electroencephalograph (EEG).

For several decades, neuroscience has devoted a great deal of effort to investigating consciousness using imaging technology like magnetic resonance tomography (MRT). Scientists have used ever more sophisticated methods to search areas such as the brain stem, thalamus and cerebrum for neuronal activity and morphological structures that might help to locate the seat of consciousness.[6] But, so far, they haven't come up with a generally applicable definition of consciousness.

Each different theoretical approach and area of experience gives us a different slant on what consciousness is. But our subjective impression is quite clear: we experience ourselves as individuals with our own ego. We believe we have free will. We are convinced that it is our consciousness that allows us to make decisions, act on them and take responsibility for our actions. Perhaps we will come to a deeper understanding of consciousness if we take a look at the unconscious. What links and what separates the two?

The interplay of the unconscious and the conscious mind

The significance of the unconscious is hugely underestimated. For Sigmund Freud it was the realm of drives, dreams and repressed memories. But the unconscious is responsible for much more than this (see Chapter 3): it actually determines many of our thoughts, feelings and actions.[7] The main reason we underestimate the unconscious is that – by definition – it mostly eludes our powers of imagination. But, even so, we can now do much more than just speculate about it.

As we have seen, our ideas and concepts all come from the emotional, sensory and motor experiences we have during our lives. The last of these, together with our innate and learned motor abilities, form the foundation of the unconscious mind. Motor functions, from the simplest reflex like the knee-jerk reaction to the complicated sequences of movement involved in walking and grasping, and skills like writing and drawing, have their roots in the unconscious (see Chapter 3). If we have driven home from work along a familiar road, we will struggle to recall the vehicles and people we passed on the way. Steering, changing gear and braking, judging distances and speeds – for the most part, we do all these things unconsciously. We get caught up in our thoughts, remembering what we have to do when we get home, while the unconscious keeps us on the right road. But if we have to follow signs for a diversion, our attention is grabbed at once. Our unconscious has been watchful all the time we've been driving, and it gives us fair warning that a conscious decision is now required. When we're awake, we are always switching between conscious and unconscious thought.

Emotions, moods and vegetative brain functions such as sleep and sexuality are all buried deep in the unconscious. We are often aware of them only in isolated moments. And the unconscious holds a vast archive of memories, which we can only access in a very limited way. Occasionally they surface in dreams or are pulled into the conscious mind quite suddenly by an event or a specific stimulus, like an enticing smell that reminds you of the cake your mother always used to make for your birthday.

In our everyday lives, we believe our behaviour is founded on conscious, well-considered thoughts and actions. And sometimes it is – but more often than not our belief is mistaken. In 1979, the neurophysiologist Benjamin Libet carried out a simple but brilliant experiment, which triggered a heated debate over the role of consciousness and free will that is still going on today.[8] Libet sat his test subjects in front of a kind of clock face on which a point of light moved around a circular scale. The subject had to choose a particular point on the scale and, once the light reached that point, he would move his finger. His brain waves (EEG) and muscle movements (EMG) were measured during the test. This meant that the researchers could measure the exact timing of the decision to move the finger, and the point at which the decision was consciously acted upon, along with the point at which the movement was prepared in the motor cortex. With the help of the EEG the researchers were able to prove what Libet called a 'readiness potential'. The surprising result, which has since been confirmed by several other research groups, was that the movement was already being prepared in the brain at a point when the test subject himself did not yet intend to move his hand. The readiness potential in the motor cortex announced the intention to act 500 milliseconds or more before the actual action. So do we lack free will? That would only be the case if we didn't view the unconscious as part

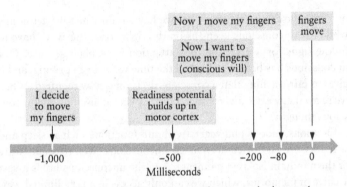

Figure 6.1: The Libet free will experiment. The timescale shows what happens up to the point when the fingers are moved.

of our own personality, and falsely assumed that our unconscious couldn't make sensible decisions or carry out meaningful actions. And this cannot be the case when all animals, guided by their unconscious, behave 'sensibly' enough to lead successful lives.

The unconscious isn't a chaotic recess of the mind (see Chapter 2). On the contrary, it is highly structured, composed of multi-modal networks and attuned to the individual's basic needs and interests. It is relentlessly engaged in maintaining areas as diverse as waking and sleeping patterns, motor functions and social behaviour. The unconscious doesn't just prepare decisions, as the Libet experiment shows; it is always *thinking* – often weighing up countless things that our conscious mind has no idea about. If, for example, we impress the person we're talking to with a clever comeback, we haven't usually thought it out consciously; the unconscious just hands it to us fully formed. We occasionally surprise ourselves with the more or less clever things we have just come out with. We shouldn't continue to let our lack of knowledge about the organizational structure of the unconscious mislead us into underestimating it so unfairly. All of our conscious thoughts and actions have their roots in the unconscious.

Reading novels, watching plays and listening to music can give us an idea of how rich this source of abilities, experiences, memories and knowledge must be. The mental images prompted by art range from exciting to repulsive or incredibly beautiful, and come with a broad spectrum of emotions. In the unconscious, they stir memories of things long past and longed for, fears and feelings of happiness. The enormous creative power stored in the unconscious is revealed in great works of art and science. Important discoveries often aren't the result of conscious, rational deductions, but of intuition – a kind of understanding that comes from the unconscious. Inspiration frequently strikes not in the laboratory or at the desk, but on a stroll or under the shower. According to legend, the chemist August Kekulé figured out the structure of the benzene ring in his sleep. The six-sided shape of the benzene ring came to him in a dream, as the image of a snake biting its own tail. The unconscious is a fund of creativity; in a profound sense, it is where our creative potential resides.

How ideas become conscious

The roots of the unconscious stretch all the way back to the early days of evolution, and the time when multicellular organisms first developed a primitive nervous system. Consciousness probably only developed when the brain reached its current size around 200,000 years ago. At this point, the high level of networking between the brain regions created the conditions for conscious thought and action.

In child development, too, everything begins in the unconscious. We don't know whether children have a kind of consciousness in the first year of their lives. We can prove the existence of an early form of consciousness at the age of 18–24 months, when children start to see themselves as independent beings (see Chapter 5). At four years old, they gradually realize that everyone has their own feelings, thoughts and intentions (see discussion of the 'theory of mind' in Chapter 5). As they begin to think of themselves as independent beings, children also gain an awareness of their own 'inner life'. They are increasingly aware of their thoughts and actions; they understand their own individuality and at the same time start to see the diversity in their social environment.

The illustration below provides an example of how symbolic ideas and linguistic terms develop in the unconscious during early childhood and gradually enter consciousness. The starting point is the experiences a small child has during a family meal. She becomes familiar

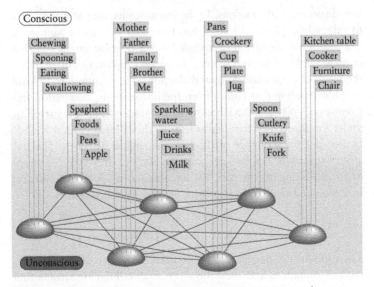

Figure 6.2: Symbolic ideas and linguistic terms, unconscious and conscious. The child's experiences at the dining table give rise to symbolic ideas and linguistic terms in the unconscious, which are hierarchically networked and can rise into the conscious mind.

with objects such as spoons and plates, foods such as bread and spaghetti, and activities such as cooking, eating and drinking. These experiences give rise to ideas, for which her parents and siblings use words, and the child learns to connect her ideas to the relevant linguistic terms. After a while, she also understands that specific terms such as 'knife', 'fork' and 'spoon' can fall under an umbrella term like 'cutlery'.

In childhood, and on into adulthood, ever larger networks of symbolic ideas and linguistic terms, ways of thinking and behaving, experiences and memories are formed in the unconscious. And, as these networks form, more and more symbolic ideas and linguistic terms become accessible to the conscious mind. We still don't have a convincing explanation for this incredible process. We have to make do with metaphors: we might think of consciousness as a kind of screen on to which items from the unconscious are projected. Things don't appear on the screen at random: they're determined by our needs and current interests. An upcoming train trip triggers a flood of considerations: looking at the timetable, buying tickets, packing a suitcase, watering the flowers. All the while, the unconscious is working behind the scenes, releasing emotions like excitement and stress. The more we think and act consciously, the more strongly we experience an autonomous sense of self. We are convinced that we're our own masters, with a high degree of free will and the liberty to make choices and decisions. But the self isn't just concerned with autonomy: it is much more focused on realizing our individuality, which is rooted in the unconscious, where our basic needs and competencies lie (see Chapters 4 and 8).

We may not know exactly how the conscious and unconscious work, but we can assume that, without a highly structured unconscious, there would be no consciousness. We might doubt whether a conscious mind could exist without a single idea. Consciousness can perhaps best be understood as a special function of the unconscious. The state of consciousness in which we currently find ourselves is largely determined by our basic needs and the situation we are in. We might very well argue that the unconscious, not the conscious mind, is the essential core of our personality that makes every individual unique. And this means we should hunt for the maestro – for our true self – not in the conscious mind, but in the unconscious.

We are still left with the question of why humans developed consciousness in the first place, when all living things had managed without it for hundreds of millions of years. It is very likely that this is one more instance of the process that drives evolution more generally: we become conscious of those things it has proven useful to become conscious of in the past. Useful in the sense that in some key areas of life (though by no means in all areas) conscious thought and actions were more effective than unconscious behaviour, and improved humans' chances of survival. The socio-cultural evolution of humans (see Chapter 1) tells us just how successful conscious thought and actions have been. And yet the overwhelming majority of our thoughts, feelings and actions remain where they are most useful to us: in the unconscious.

FUNDAMENTALS FOR THE FIT PRINCIPLE

Everyone has their own ideas

> The word belongs half to him who speaks it, and half to him who listens.
>
> Michel de Montaigne (1533–92)[9]

Our words often aren't understood exactly as we meant them. Some ideas and linguistic terms have a different meaning for each of us, resting as they do on the experiences we have had, particularly in childhood. Let's take the example of moral development: children develop different ideas of morality depending on how their parents and carers bring them up. In addition to this, the experiences that children have are always selective and contingent on their basic needs and competencies, their emotional state and their current situation. Even siblings growing up in the same family have different genotypes and therefore internalize different experiences and develop their own ideas (see Chapters 2 and 3).

Ideas are so often understood differently because they hardly ever appear in isolation. Within the unconscious, they are networked into a great many other ideas, each of which has an influence on the others. The word 'learn' is connected to other terms, including 'by heart', 'diligence' and 'grades'. In conversation, these networks of ideas mean we may start to ramble or digress. They are also the reason behind a phenomenon we encounter again and again: ideas are not easy to change – even with the best arguments. Changing an idea for good means adapting the ideas networked around it as well, and this is something that can't be achieved through rational insight alone: it requires additional experiences.

The ideas we bring up in conversation or put down on paper are always influenced by our current situation in life. How we feel also plays a role: whether we are well-rested or dog-tired, sober or tipsy. But ideas are mostly determined by our profile of basic needs. We aren't usually conscious of the influence our basic needs have on us. Take Karl: as a teenager, he joins a local evangelical church. He fiercely defends its rigid religious ideas – but not because he's completely convinced by them. He feels lonely in his private life and the church community accepts him. The sermons give him a clear direction and advice on how to live his life. As a member of the choir, he gets to sing on stage in front of hundreds of believers and receives recognition for it.

Most ideas and beliefs aren't fixed for all time. They change over the course of our lives based on the experiences we have. After ten years, Karl leaves the church: he has now found emotional security with a partner, started a family and discovered that he can manage his life independently. Political positions can also change over time – as the common saying goes: 'If a man is not a socialist by the time he is twenty, he has no heart. If he is not a conservative by the time he is forty, he has no brain.'

So how should we treat our own ideas and those of other people? While he was a member of the church, it would have been impossible to change Karl's convictions using rational arguments. We could have called his church and its religious promises a sham and dismissed his enthusiasm as blindness, but he still wouldn't have changed his mind. We understand him in hindsight, though, when we realize why he found the church helpful and why he subsequently left. His life had changed to such an extent that he was able to satisfy his basic needs more fully in another way. Now he, too, is prepared to doubt the church's religious message.

We don't often manage to convince someone that our ideas are right – but that doesn't stop us trying over and over again. And the attempt can sometimes cause misunderstandings, which in the worst case may damage our relationship with someone. We can co-operate and form friendships much more successfully when each of us understands the individual experiences that we and other people have had. But what is really helpful to us is when we understand the needs, interests and circumstances that gave rise to our ideas and words, as well as those of others. And because we often know too little about these things, we should grant our fellow humans the right to their own ideas as much as possible, treating them with respect and always questioning the origin of our own ideas, particularly in relation to our basic needs.

There are those rare moments when we sense that another person understands us completely. We take joy in these moments not because we have reached a consensus, but because we feel like we have found a kindred spirit.

Why collective ideas are so powerful

... words are, of course, the most powerful drug used by mankind.
Rudyard Kipling (1865–1936)[10]

There are collective ideas that are so powerful they give rise to ideologies, which in turn can start mass movements and lead to the formation of governments, and often also spark conflict and war. National Socialism was one such ideology, spawning Hitler's German *Reich*; or Communism, on the dogmas of which Lenin founded the Soviet Union and Mao built modern China. The power inherent in such ideologies can be seen in the horrific destruction they caused across the world in the twentieth century. The populations of communist and fascist countries, and many others, were subjected to a reign of terror. Many millions of people were ostracized or killed. Millions of people were manipulated and incited to kill. Today, fundamentalist groups with pseudo-religious ideas are terrorizing people in the countries of the Middle East, and increasingly also in the West. What makes ideologies so powerful?

Ideologies are still wrongly viewed as the driving force behind mass movements such as the October Revolution. But Vladimir Ilyich Lenin didn't manage to win the Russian people over to his cause in 1917 because they were convinced by his ideas. Only a vanishingly small minority had read *Das Kapital*. The social and economic ills of the time were what put Lenin in power. People believed the Revolution would help them escape poverty and unemployment. They were hoping for social justice, better education and decent healthcare.

In order to secure power and exercise it, ideologies abuse humans' boundless desire to have their basic needs satisfied. They convince people to live with repressive regimes and accept tremendous injustice by always dangling in front of them the hope that their living conditions will improve. Some ideologies use this strategy to develop a destructive dynamic which, in the worst case, can last for decades and cause great human suffering. Ideologies lose their power when people realize that the things they have been promised will never materialize and their living conditions aren't going to improve. The population of the Eastern Bloc countries came to this conclusion in the 1980s. In East Germany it wasn't just the extreme surveillance, corruption and abuse of power that made citizens lose faith in the government. People realized they would never attain the prosperity and social freedoms that West Germany enjoyed and, worse, that their country had fallen so far behind it was on the point of collapse. It was then that their belief in the East German brand of socialism finally crumbled – and that was what ultimately brought down the Berlin Wall.

Some countries, including Switzerland, were spared from fascism and communism in the twentieth century, but not because their populations were particularly resistant to ideologies. They were protected by material prosperity, existential security and social justice. If living conditions deteriorate – and I sincerely hope they don't – then we can't rule out the possibility that even the upright citizens of the Swiss Confederation might fall prey to an ideology and fall into line behind a *Führer*. The power of ideologies lies in the implicit promise of glory they make to people living in difficult circumstances. The fertile soil in which ideologies germinate is composed of unsatisfied basic human needs and the hope of a life worth living.

Ideologies ultimately disappear when they can't fulfil their promises (such as work for all) and, more importantly, when they don't manage to satisfy people's basic needs. Religions such as Christianity are quite a different matter. They have survived for thousands of years because, unlike ideologies, they don't aim to make the world a better place. They have always seen the misery of the world as part of mankind's lot. For all the moral aberrations to which religions have fallen prey (the sale of indulgences, for example), they have remained credible because they genuinely help people satisfy their basic needs. They give people emotional security and hope, and provide a sense of community and solidarity. And they take special care of the weak and the disadvantaged. St Francis of Assisi, a shining light of Christianity, expressed this basic attitude when he said: 'Blessed is the man who bears with his neighbour according to the frailty of his nature as much as he would wish to be borne with by him if he should be in a like case.' Today, some of the important work performed by religions over thousands of years has been taken over by the state and the economy. State institutions support unemployed people by providing advice centres and financial aid, and assist elderly people who would otherwise be living in poverty. However, the core objective of religions, to accept and support people in all their strengths and weaknesses, remains. It is also a fundamental consideration for a fitting life.

7

From Nature to the Man-made Environment

In order to survive, all living things need not just any environment, but one suited to their needs.

> *In living nature nothing happens that does not stand in a relationship to the whole.*
>
> Johann Wolfgang von Goethe (1749–1832)[1]

The environment as such doesn't exist. Each living thing has its own environment, to which it has adapted over the course of evolution. In order to survive, plants and animals therefore need not just any environment, but one suited to their needs. Our ancestors became an exception to this rule very early on. They managed to survive in a huge range of environments, from the desert to the jungle or frozen forest. Over the past 10,000 years humans have come up with better and better ways of ensuring that their basic needs are met, starting with overcoming periods of famine by laying down stores (see Chapter 1). Today we are capable of surviving in environments extremely hostile to life: on a submarine in the Mariana Trench, the earth's deepest point 11,000 metres under the sea, or on the International Space Station, 400 kilometres outside the Earth's atmosphere.

Socio-cultural evolution has turned us into a great exception among living things. Our drive to understand and control the world has expelled us – in a biblical sense – from nature (see Chapter 1). These days, very few people live in *Homo sapiens*' 'natural' habitat, a small, familiar area of the natural world. The social and economic living space we have created spans the globe, and we are finding its social, cultural and economic complexity increasingly opaque and overwhelming.

The Anthropocene[2] is a new age in which humans have become one of the most significant influences on the earth's biological, geological and atmospheric processes.

Despite the fact that we humans are extremely adaptable, we also still depend on certain environmental conditions that prevailed during the evolution of *Homo sapiens* and influenced our genotype. Our existential, social, physical and psychological needs are just the same as they were 100,000 years ago. But the environment we live in today is so different from that of our forefathers that it fits our basic needs less and less. In Chapter 7 we will try to answer the following questions:

To what degree has scientific, technological and economic progress changed our environment? How much has our relationship with nature, and therefore our quality of life, been damaged? What kind of environment have we created for ourselves? How much has it changed the way people live together? Do we feel that anonymous state and economic institutions have taken control away from us as individuals? Why does the modern environment fit our basic needs less and less, and how does this affect our mental and physical wellbeing?

ESTRANGED FROM NATURE

For 200,000 years our ancestors lived in and off the natural world. They only entered their caves or huts to sleep or take refuge from bad weather and enemies. They spent most of their time outside, engaged in activities such as hunting and gathering plants (see Chapter 1). Humans' close ties to nature were what enabled them to survive. Over the last 12,000 years humans have moved away from nature, first into small, modest settlements, then into towns and finally megacities.

The process of urbanization has seen exponential growth in the past 200 years. At the start of the nineteenth century the vast majority of the world's population still lived in rural areas. By 2008, more people were living in cities than in the countryside, and by 2030 the UN estimates that between 5 billion and 8 billion people worldwide will spend their lives in a metropolis.

How does this estrangement from the natural world affect our well-being? Nature has largely been lost from our day-to-day lives. Most people spend only a fraction of their time in nature – if any at all. They might jog in their spare time, go on a safari holiday in Namibia or take a boat to Svalbard. We can feel how important nature is to our wellbeing when we go out into the woods, climb a mountain or walk along a river. We enjoy being in the natural world and feel its restorative power. Nature still stirs a sense of deep familiarity within us. Going to a new city feels strange and can be stressful, whereas a part of the natural world we don't know is still somehow familiar and has a calming effect on us.

Children have a particularly close relationship with nature. Small children love splashing about in puddles and wading through mud. They enjoy picking berries and collecting pine cones or snail shells. Older children love swimming and diving in lakes or damming streams. They play hide and seek and chase each other through the woods. They cut

Figure 7.1: From caves to megacities.

sticks and have battles with them. They're fascinated by matches and love lighting campfires. Their games have an archaic quality, as if they were playing in a long-ago era.[3] However many toys they have, few children prefer being in their bedrooms to playing outside in natural surroundings with their friends. Nature has an incredibly strong pull for children. It awakens 100,000-year-old learning impulses, prompting them to have the experiences they need for their development.

People may be living in the midst of nature less and less, but their interest in it has steadily increased. They want to understand nature better in order to make better use of it. In the last 200 years, we have heaped up an immense store of knowledge about the natural world. Our ancestors didn't know there were such things as bacteria. Now, we not only know that certain bacteria like staphylococcus can cause serious infections, but have also developed antibiotics to combat them. We have learned that many kinds of bacteria have positive attributes and can even be vital to keeping us alive. Our gut relies on lactic acid bacteria that support the digestion of food. Countless other bacteria live in similar symbiotic relationships with plants and animals. All these discoveries about the essence of nature have brought us to the conclusion that bacteria, like all forms of life (including humans), are part of a single, global ecosystem. Any damage we do – both on a small scale, like harming bacteria with fertilizer and chemicals, and on a large scale, by causing the extinction of plant and animal species – will also have a negative effect on us. The famous entomologist Edward Wilson put it this way: 'If there were no more insects, I doubt that mankind would survive more than a few months.' Without exaggerating, we might add: if there were no more bacteria, mankind would die out in a matter of weeks, along with the rest of the world's flora and fauna.

We have done plenty of damage to the natural world. The over-use of resources like ore, coal and rare minerals, and an industrialized economy, exploiting palm-oil plantations, for example, have become a monstrous kraken with the natural world in its grip. The release of CO_2 is damaging the atmosphere, the rainforests are being chopped down and the biosphere is being destroyed. We are poisoning the seas with chemicals and clogging them with plastic waste. Our relationship with nature has become exploitative, and we now fear the consequences of our carelessness and excessive pursuit of profit.

We know we have to end our destruction of nature at once, and not just because we rely on it for our own existence. We must also give some thought to the harmful consequences for our wellbeing if we could no longer live in nature at all. Could it be that certain psychological conditions such as ADHD stem from the fact that we have become estranged from our original way of life in nature? Hyperactive children don't seem unusual when they're out in the woods. Can we (and children in particular) manage without the natural world in which we lived for hundreds of thousands of years, without causing ourselves mental and physical harm?

A high-achieving society with a hollow culture

There are good reasons for increasing numbers of people moving to towns and cities. They benefit from a better quality of life and higher, better-protected incomes. They live in comfort, with great material wealth – some people seem almost to be drowning in a flood of consumer goods. The enormous progress that has been made in healthcare has caused a dramatic decline in child mortality and a continuous rise in life expectancy. In Europe, the child mortality rate (deaths under the age of five for every 1,000 live births) has fallen in the last 100 years from 200 to 8. In the past twenty years, it has fallen by almost 50 per cent worldwide. In 1860, average life expectancy in Germany was forty-one for women and thirty-five for men. By 2015, it had risen to eighty-three and seventy-eight respectively, and is set to increase further in the decades to come. These days it isn't a high mortality rate that worries people in Europe, but overcrowding and an ageing population. Their greatest concerns are no longer famine and malnutrition, as they were in the nineteenth century, but overeating (obesity) and its negative consequences, which include diabetes and heart disease. The social system is so highly developed that even people in difficult existential circumstances can still live with dignity. The frequency of violence, be it in the form of wars or criminality, has fallen globally – even if that seems hard to believe when we think about the conflicts currently raging all over the world. One thing is clear: when living conditions improve, levels of violence fall.[4]

Education is another great achievement of the modern age. In the past, bringing up children was a task for families and the small communities they lived in. Children developed by living with people of different ages who served as role models. They learned the local language, the rules of social interaction and the values they lived by from the people around them. They acquired abilities and knowledge by being included in the adults' activities, such as collecting berries or wood, hunting or tending livestock. Children were present at celebrations from a very young age, at first just watching and listening, and then singing and dancing with the others and internalizing the practices and traditions of their community.

Education in the form of systematic schooling with a focus on academic abilities like reading, writing, maths and the communication of cultural achievements began around 4,000 years ago, but for a long time it remained the preserve of a small elite. The primary school, a school for all children, was introduced in the nineteenth century in the wake of industrialization. Society recognized that scientific, technological and economic progress had made it necessary to raise the level of education in the general population. Today, we have a very sophisticated education system that takes us from nursery school to the completion of a university degree, and continues with adult education and training up to retirement age and beyond. In the past 150 years, this development has led to a huge transfer of responsibility, as the tasks of bringing up and educating children have been passed from families and small communities to state-run institutions.

Many countries still don't have compulsory education, and their education systems are underdeveloped. But we should feel encouraged by the great progress that has been made across the world in the last few decades. Illiteracy is in decline worldwide.[5] The idea that everyone has a right to an education has a good chance of becoming a reality. And that doesn't just go for boys: increasingly, girls are gaining unrestricted access to education, even in developing countries. In some parts of Europe, women are already educated to the same level as men, or an even higher level. In Switzerland, 60 per cent of girls, but only 40 per cent of boys, attend a grammar school.

Like education, culture should help to give life meaning and contribute to community cohesion. But the globalization that has taken place in many areas of life has now led to culture being drained of meaning, a fact that the cultural critic Neil Postman lamented as early as the 1980s.[6] In our ancestors' communities, people of all ages participated in cultural activities. They told each other stories, sang together, danced and made music. They loved eating and celebrating together. They created shared values and symbols; they had their own morality and their own local heroes. Their culture served to strengthen social and emotional cohesion and create a shared identity, through rituals, traditions and values. The transformation of small communities into anonymous societies has brought with it a weakening of cultural as well as social cohesion. Rituals, practices and traditions are becoming increasingly rare, and values are shared less and less. Active participation in culture is increasingly being replaced by a national (and, more recently, global) entertainment industry, the output of which we tend to consume passively. Billions of people spend their evenings sitting alone or in pairs in front of the television or the computer, being entertained. Daily soap operas tell stories of love and loss, intrigues at home and at work, sickness and death, in ever-changing scenarios. All the emotions people are missing so desperately in their own lives are played out for them in the comfort of their living room. When the evening is over, we are left with a stale sense of hours wasted: even the best of all virtual worlds can't replace a real community.

Of course, we still go out and enjoy wonderful concerts, captivating plays and stunning art exhibitions. But these events no longer fulfil the original task of culture – namely to create space for an active communal experience. Cultural property, practices and traditions, stories, songs and dances used to be guarded like a precious treasure and handed down from one generation to the next. Today, trends in the visual arts, film, music and literature change from week to week all around the world.

A gigantic, profit-oriented entertainment industry may keep people happy to some degree, but it contributes nothing to cohesion in an anonymous mass society. In a small community, culture is a kind of interpersonal glue, and experiencing it gives people's lives meaning.

In Chapter 10, we will look at how we might find our way back to a community-based culture.

THE TRANSFORMATION OF OUR SOCIAL ENVIRONMENT

Humans are deeply social creatures. Long term, we can only exist within a reliable network of relationships. For at least 200,000 years this network consisted of family clans and tribe-like communities (see Chapter 1).

Our ancestors spent their whole lives in a network of a few hundred people with whom they were very familiar. They depended on each other in an existential sense, since their basic needs could only be satisfied within their family and community. And this meant they developed a kind of cooperative behaviour unique among living things, which we have retained to this day.[7] But within the space of a few generations, modest communities based in small areas have become an anonymous mass society. Our evolutionary heritage makes us ill-suited to this social environment. Our wellbeing depends on a network of relationships with familiar people.

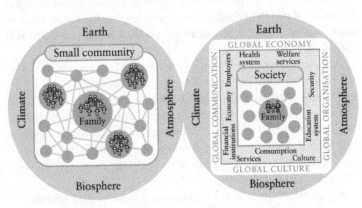

Figure 7.2: From families and small communities (*left*) to mass society, where state institutions and economy are surrounded by global organizations (*right*).

From extended families to nuclear and blended families

The family is celebrating Eva's third birthday. Her grandparents, uncles, aunts and cousins have travelled to be there. Eva is unwrapping her presents when she suddenly runs after a child who has 'stolen' a present from her. She trips and falls, and starts to cry. No one but her mother can comfort her.

The first thing children do when they're upset is seek support and protection from their mother and father. Parents are every child's most important attachment figures. But children also want trusting relationships with people from their extended family and their community. In the past, a broad support network was tremendously important for children's survival, because parents, in particular mothers, often died young. The importance of an extended network of carers can also be seen from the fact that children don't just bond with their biological parents: they will bond with any adult who cares for them enough. The strong bond between children, parents and carers involves a higher level of trust than any other kind of relationship (see Chapters 4 and 5).[8] It is what has been keeping families together for 200,000 years.

As society and the economy have developed over the past 150 years, the family has gradually lost the emotional and social stability it once had. A range of factors has contributed to this destabilization, one of the most significant being the long-overdue emancipation of women. Crucial to this was the invention of the contraceptive pill, the social repercussions of which are still massively underestimated. For the first time in human history, women don't have to get pregnant when fate decides they should. They can choose whether and when they want to have children. The concept of equal rights for men and women has been another important factor in women's emancipation. Equality has been achieved to a large extent in the education system, to some extent in society, and rather less in business. The decline of manufacturing industry, and with it the reduction in physical labour that had been largely carried out by men, was also crucial – as is the rise of the service economy, in which women's social and linguistic competencies are particularly sought after. This multilayered development has permanently changed our behaviour in romantic relationships and our understanding of gender roles, which in turn has had a powerful effect on family life.

Transformation in the past fifty years

- Emancipation of women in terms of education, jobs, and status in the family and society
- Contraceptive pill and artificial insemination
- Families with two working parents
- A new understanding of men's and women's roles
- Increased mobility

Effects on living together

The transformation of family life and its stresses and strains

- Nuclear families with one or two children
- Mothers and fathers being put under increasing pressure
- High divorce rate
- Parents and partners living apart
- Single-parent families
- Blended families with step-parents and stepchildren
- Increased mobility leading to fragmentation of living space
- Increased stress and insecurity for both parents and children
- Lack of support and solidarity from relatives and communities
- More dependence on society, e.g. the welfare state

Never before has there been such a radical change in family structures. It comes as no surprise, then, that parents aren't the only ones finding their everyday lives difficult: society as a whole is, too. Large families with lots of children and relatives have shrunk to become nuclear families with one or two children, whose relationships with other relatives are more distant. The divorce rate in Western countries has risen from 5 to 50 per cent in just a few decades. Partners and parents increasingly live apart. In Germany, around 20 per cent of children grow up in 'post-divorce' families. Half of these children are brought up by single parents, most often mothers. The other half live in various different arrangements, including blended or stepfamilies.

The number of people who want to have children together without getting married is on the rise, as is the number of people who don't want children at all.

There has been a significant decrease in the stability and continuity of care that is so important for children's wellbeing. The number of trusted attachment figures that children have has shrunk, and the number of people looking after children they don't know well has grown. This is not just the case in a family setting, with childminders and au pairs: it's also due to the constant turnover of carers and teachers in nurseries and schools. The fragmentation of everyday life in terms of both time and space, which is mainly a result of our increased mobility, has a negative impact on quality of life for both adults and children. Children are missing out on long-term relationships with attachment figures and, even more importantly, with other children. They are finding it harder to enter into and maintain friendships.

Can this be good for us in the long run? We may well doubt it. There are clear indications that increasing numbers of children are emotionally and socially stressed. Young people and even children have recently started to be diagnosed with burnout for the first time.[9] We should be concerned that some children will grow up to be insecure adults, having failed to build up a basic trust in someone loving and caring for them unconditionally. If we want our children to become people with a healthy level of basic trust, then we need to rethink the spaces in which they live. On their own, nuclear families can't provide continuity and stability in childcare. This requires a reliable community, along with a society and an economy that demonstrates an appropriate level of understanding for families and gives them enough support (see Chapter 10).

But children aren't the only ones who need emotional security and affection – adults need much more of these things than is widely assumed. More and more people in the younger and middle generations are suffering from emotional insecurity and loneliness. Today, partnerships tend to last for a limited time and crack more easily under pressure – a fact reflected in the high divorce rate. Emotional and social insecurity increases further as we age. Relationships with friends and relatives become weaker and finally dissolve altogether. As they enter retirement, people also lose a lot of their social status;

they feel as if they are a burden on their family and the state, which strikes an additional blow to their wellbeing and sense of self-worth.

From an evolutionary point of view, humans are made for life in a community. People of all ages want to feel emotionally secure and to have reliable relationships with trusted people. They want to live in social surroundings that give them a sense of belonging, and to have an influence on the people around them. Chapter 10 will explore what new forms of communal living might look like.

From small communities to mass society

The train carriage is quiet. People are travelling home from work in the evening, and they're tired. They are reading the paper, playing on their phones, looking out of the window. Suddenly, the train stops. There's a power failure. People wait, growing increasingly impatient and angry. Eventually, they start to talk to each other and consider their options for getting home. Should they walk? Take the bus? Can anyone come out and give them a lift? Solutions are quickly found, and at once everyone is in a good mood. The next morning, they get back on the train and say hello to their new acquaintances.

Needs create relationships. This was as true for our ancestors as it is today. In small communities, people were extremely dependent on one another in emotional, social and existential terms. Alongside the family, the small community was once the second-prevalent form of communal life (see Chapter 1). It was made up of not more than a few hundred people, who all knew each other and lived together in various social groups. The community shared a language, customs and religious ideas. People came together to find food, take care of children, or protect people and property from violence, often under difficult conditions, and this work knitted communities together very tightly. The division of labour was also important for creating a sense of cohesion. Each member of the community (ideally) used their talents and knowledge to benefit everyone, and profited from the competencies and achievements of others. They were able to make use of their strengths and receive support for their weaknesses, and felt they were in good hands. The roles they were assigned and the social status they held in the community constituted a large part of their identity. Such a tight web of mutual aid and dependency almost certainly caused some conflicts and violence, but it also provided a strong sense of existential, social and emotional security.

Small communities were widespread in the western hemisphere until well into the twentieth century. In many countries, billions of people still live in these communities. They sit together in the evenings, tell each other what has happened over the day, and pass on news and stories. They sing and dance with each other, and hold sprawling celebrations of important events such as births and marriages. They laugh, cry and argue. Like families, small communities are not free from conflict. But they give people the reassuring certainty that they will always have this group of familiar people to rely on. In her book *The Village Effect*, the psychologist Susan Pinker describes communal life in a mountain village in Sardinia. In her view, lifelong trusting and stable relationships are vitally important to us.[10]

In the modern age, these communities have grown into ever larger social units. The nineteenth century saw the founding of nation-states, which governed people's lives in a whole new way. In the second half of the twentieth century, international structures like the EU and global organizations like the UN and the WHO came into being. A similar thing happened in the world of business. The Industrial Revolution saw the founding of countless regional firms, which gradually joined together. These larger organizations went on to become transnational companies and then global empires such as ExxonMobil and Nestlé.

We can see just how profound this social transformation was from the changes in how communities have been regulated. In the past, laws were often passed down by word of mouth and were binding for a group of just a few hundred people. Within 200 years, these laws developed into written constitutions and legal systems that were binding for millions of people, and finally into a universal understanding of law to be implemented worldwide. Today, the Universal Declaration of Human Rights and the Charter of the United Nations are accessible to everyone and can be downloaded from the internet at any time. This rapid development has taken place beyond the legal system as well, in all areas of society including healthcare and social and communications systems. In our highly complex modern society, huge, anonymous state and business institutions have taken over the tasks that used to be performed by a few hundred people in a small community.

FUNDAMENTALS FOR THE FIT PRINCIPLE

It was a wake-up call. In 1972, the Club of Rome published its report *Limits of Growth*. Since then, we have become increasingly aware of the threat that our mania for growth poses to the natural world. It is gradually dawning on us that our man-made environment not only damages nature, but is also less and less well suited to humans.

Relationships with trusted people are indispensable

I tell ya, a guy gets too lonely an' he gets sick.

John Steinbeck (1902–68)[11]

We still live in family-like structures and communities, though they have shrunk drastically in comparison with earlier times. Our network of relationships is often split into the different areas of our lives: partners, parents, work and free time, in which we can only ever satisfy some of our needs and carry out a limited range of activities. The circle of people with whom we are truly familiar and have long-term or even lifelong relationships has become very small. We live in an anonymous society composed of many millions of people, with whom our relationships are superficial or non-existent. Many of us feel stressed by the constant interaction with people we don't know, at work or when we're out shopping, for instance. We rely on other people to satisfy our basic needs much less than our ancestors did. And on top of this we live in a society where we are in constant competition with each other, always having to prove our worth as partners and employees, and in constant danger of dropping out of all our relationships and becoming socially isolated. More and more adults of all ages are suffering from a lack of emotional security, affection, social recognition and a sense of belonging. Small communities have

Figure 7.3: Conviviality and loneliness. (*top*) Rural celebration in the Netherlands (Pieter Brueghel the Younger); (*above*) high-rises in Hong Kong.

the power to provide care and solidarity and discourage competition and stress, and this power has been lost in an anonymous society where we have to be largely self-reliant. A frequent complaint is that our extreme individualism has come at the cost of solidarity.

We lead our lives as if we could manage without steady, sustainable, trusting relationships. But these relationships, along with cooperative communal living, are our only source of emotional security and social esteem. In the future, we must give back to the family and the community – albeit in a new form – the significance that is essential if we are going to satisfy our emotional and social basic needs (see Chapter 10).

Self-determination and personal responsibility

There is only one problem for this prison, namely that of proving that it is no prison, but a bulwark of freedom . . .

Friedrich Dürrenmatt (1921–90)[12]

In the space of a mere sixty years, individual freedom – the opportunity to make decisions and choices for ourselves – has increased to undreamed-of proportions. Many of us are able to choose what career we go into and where we want to live and work. We can pick from a broad range of consumer goods. We no longer buy our bread from the only baker in the village, but in any one of a large number of supermarkets. We don't have our clothes made by a tailor: we order them from one of the countless companies operating online. We can choose which doctor, hairdresser or tax adviser to see. We work, sleep and get our entertainment in a huge number of different places. We can take the high-speed train to see an opera in Paris or a fashion show in Milan. We can fly to London for a shopping trip or the Maldives for a holiday. Our mobility has increased to a degree that is damaging the environment. Volatility is one of the keywords of our time. And it doesn't just apply to the stock market, but to human lifestyles.

We pay a high price for these freedoms. We have become dependent on the anonymous institutions of the state and the private sector. Very few people still manage to be self-sufficient in terms of food, energy and living space. We work for a salary and use it to buy food and accommodation, to keep us mobile and gain access to a host of services and a globalized consumer world. State institutions are with us all our lives, from nurseries and schools to unemployment benefits and the health service, and eventually retirement homes. We are well looked after, but we have lost control over essential areas of our lives. We are sitting in a comfortably furnished cage. The anonymous institutions that provide for us are our jailers. They make it easy to satisfy our basic needs, which was a heavy burden for our ancestors, but which also contributed to the meaning of life (see Chapter 4). Existential challenges have dwindled to become competitive achievement, the pursuit of profit and the drive to consume. More and more people are feeling overstretched. They sense their lives are no longer under their control, and they are looking for alternative ways of living. They are pursuing greater independence and are eager to take on more responsibility. They are founding self-governing communes, for example, where families with children, single people and older people all live together (see Chapter 10).

People don't just feel their lives are out of their control: they are also losing faith in the state and the economy. The financial crash of 2008 gave people a sense of insecurity that still hasn't abated. They are afraid of mass unemployment and a general decline of affluent society. Their insecurity is only reinforced by the fact that society and the economy have become opaque to them. They wonder who is really responsible for the state and private-sector institutions that provide food and energy. They doubt that the government and the CEOs are really as competent and far-sighted as they claim to be. Even politicians and business leaders are increasingly taking the view that our successful democracy and our booming market economy are in crisis.

There is an apocalyptic mood in the air. People sense that things can't go on like this, but they have no idea how society and the economy need to be structured in the future. They feel helpless in the same way that people did at the start of the nineteenth century after the French Revolution. They simply couldn't imagine what their world might look like once the feudal system had been abolished. Power and land had been in the hands of the aristocracy and the Church since the ninth century, and serfdom was widespread. But in the space of a few decades (with some temporary setbacks) the first democratic republics emerged, and, with the arrival of industrialization and the service economy, completely new economic structures developed. Now, 200 years later, we are due another social and economic reform, which may well go even deeper than the transformation that followed the French Revolution. The people of that time couldn't imagine representative democracy and the free market economy. We too find it difficult to picture what society and the economy might look like in the future (see Chapter 10).

I am confident that future generations will manage to transform our society and economy in such a way that people can better satisfy their basic needs. 'Liberty, equality and fraternity' was the slogan of the French Revolution. Ours might be: 'Solidarity in our diversity'. And by that I mean a life in which people can satisfy their basic needs fully and use their competencies in ways they themselves have determined. A life in which they can be part of a community that accepts them just as they are, with all their strengths and weaknesses (see Chapter 10).

8

Living the Right Life:
The Fit Principle

Being true to our individual nature is a challenge
that keeps us on our toes all our lives.

Before his end, Rabbi Sussja said: 'In the hereafter, they will
not ask me: "Why were you not Moses?" They will ask: "Why
were you not Sussja?"'

The Fit Principle builds on the age-old human need to exercise our individuality in harmony with the environment, and in so doing become fully ourselves. This – as the story of Rabbi Sussja shows – is not a new idea. Rather, the Fit Principle is an attempt to reshape old wisdom to fit today's world.

If Rabbi Sussja's recommendation was easy to follow, I wouldn't have written such a long book about it – perhaps no book at all. But exercising our individuality is probably the greatest challenge life throws at us. We want to realize our strengths, and we have to learn to deal with our weaknesses. And, however hard we try, we can't do it alone. We depend on our social environment – relatives, friends, work colleagues – to support us and respect our individuality.

In Chapter 8 we will address the following questions: what elements go to make up a person's individuality? What do basic needs, competencies and ideas contribute to individuality? What significance does the environment have for an individual? Which areas of the environment are particularly important for us, in order for us to receive social recognition, for example? How can the Fit Principle help us gain a better understanding not just of ourselves and our own situation, but of other people and their situations? And how can we succeed in exercising our individuality in harmony with our environment? What must we contribute to that harmony, and what does the environment have to contribute?

HOW THE FIT PRINCIPLE CAME ABOUT

Over the course of evolution, human development has resulted from a relentless interaction of genotype and environment. To some degree, we see that same process in microcosm as each of us strives to exercise our individuality in harmony with our environment, from childhood to old age. This was probably the greatest insight that my academic and clinical work afforded me. It was like doing a jigsaw puzzle: this one piece made all the other pieces of the Fit Principle fall into place.

The pieces of the puzzle

Numerous books by exceptional figures in the humanities and natural sciences made essential contributions to the development of the Fit Principle. One of these figures was the religious philosopher Martin Buber. He developed the 'dialogical principle' 100 years ago, based on Hasidic mysticism, Christianity and the existential philosophy of Søren Kierkegaard.[1] Buber argues that humans form their identity mainly through their relationship with the environment that surrounds them. The 'I' is born out of the encounter with a human other, the 'you' (often called the 'I–Thou relation' in English). His dialogical principle gave me a new way of seeing our relationship behaviour. Abraham Harold Maslow's work places great emphasis on the uniqueness of the individual as a *conditio sine qua non*, an indispensable condition of human nature.[2] He was one of the founders of transpersonal psychology, which I encountered as a young doctor at the University of California in the 1970s. At that time, I was particularly impressed by his hierarchical pyramid of human needs.

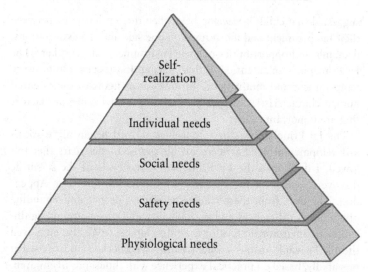

Figure 8.1: Maslow's hierarchy of needs.

At around the same time, the Chilean neurobiologists Huberto Maturana and Francisco Varela were attempting to define the characteristic organizational features of living organisms.[3] They introduced the concept of *autopoiesis*, which essentially means self-creation and self-maintenance. These two terms are not only significant parts of evolution, but have also become important elements of the Fit Principle, in particular when it comes to understanding child development. In the 1970s and 1980s the psychologist Aaron Antonovsky turned away from pathogenesis, the origin and development of illnesses (and the focus of traditional medicine), to focus on *salutogenesis*, the origin and development of health. Rather than worrying about what makes people sick, he wanted to find out what keeps them healthy.[4] The principle of *salutogenesis* was the start of an approach to medicine that concentrated not on pathology but on the healthy organism. It was an approach that chimed with my thinking as a doctor and a researcher: I had learned a great deal more from observing 'normal', healthy children than from the study of illnesses.

Stella Chess and Alexander Thomas, both active in child psychiatry, came up with the expression 'goodness of fit' in the 1980s. They suggested that children develop best when there is harmony between their temperament and motivation on one side and the expectations, demands and opportunities of their environment on the other.[5] The Fit Principle is an extension of this concept. It doesn't try to measure temperament and motivation, but looks at as many of the essential human characteristics as possible and all the environmental factors that are important to us.

The Fit Principle was first conceived when I began my work as a developmental paediatrician in the 1970s. In the years that followed, I developed the Fit Principle on the basis of the scientific discoveries made in the Zurich Longitudinal Studies (see Appendix). The data from these studies gave me a deeper understanding of children's differences and individuality and the enormous significance the environment has for their development. We also employed the Fit Principle in our clinical work. We checked and revised it repeatedly, based on practical experience with thousands of children. Over the past thirty years, we have developed a tool that has proven very helpful in areas such as sleep consultations. Sleep disorders are very common in children, and around a third of pre-school children suffer from some problem with getting to sleep or sleeping through the night.

Night-time sleeping patterns change dramatically during childhood. In the first year of life, the length of time children sleep at night almost doubles, before decreasing steadily all the way to adulthood.[6] At all ages, sleep duration varies greatly from one child to the next. One-year-old babies sleep for an average of twelve hours a night. But some get by on less than ten hours' sleep, while others need up to fourteen hours. The realization that children's need for sleep differs at every age has immediate practical consequences. The widespread opinion among experts and laymen, that the more sleep children get the better they will develop, just isn't true. Like adults, children feel well-rested when they are able to get the amount of sleep determined by their individual biological need. Too little sleep will have an impact on their cognitive function, behaviour and wellbeing. And if children are made to spend longer in bed than they can sleep for they will lie awake at night.

In our sleep consultations, we asked parents to record their child's

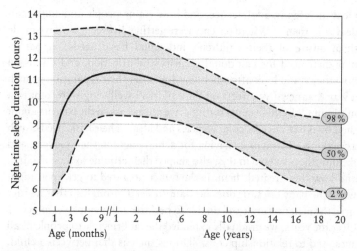

Figure 8.2: Duration of night-time sleep between the ages of one and twenty. Bold line: 50th percentile (average); upper and lower limits: 98th and 2nd percentiles; in each case 2 per cent of the study population are above or below this limit. (source: Zurich Longitudinal Studies)

sleeping patterns for ten days using a sleep chart. It emerged that one of the most frequent causes of sleep disorders (though not the only one) was parents misjudging how much sleep their child needed – usually by overestimating it. Our advice to parents was to tune into their child's individual need for sleep.[7] We found that there was huge diversity among children in all areas of development, including eating behaviour, motor function and language. Parents, carers and teachers can support children's development best by taking into account their individual profiles of basic needs and competencies (see Chapters 4–6).

Over the past thirty years we have used the Fit Principle not only with children who are experiencing developmental disorders and behavioural abnormalities, but with those who are having problems at school and finding learning difficult.[8] When advising parents and teachers, we used it to find a balance between the children's individual competencies and the demands that adults were placing on them.[9] We have been able to use the Fit Principle to help parents and teachers understand the profound changes that basic needs, competencies and ideas undergo in adolescence, and how young people and adults can deal with them.[10] Where parents were getting divorced, we have made them aware of their children's individual basic needs, in particular their need for emotional security and affection, and discussed how they could look after their children after their separation in such a way that negative effects on the children's wellbeing were kept to a minimum.[11] In these conversations, we always asked the parents about their own basic needs, competencies and ideas. Their basic needs, such as making a living and striving for achievement in their careers, and the time they devoted to this, also shaped their attitude to their children and how they treated them. Is the father prepared to get up at night when the baby cries, or does he want to sleep through so that he isn't tired the next morning and unable to perform as his employer expects? Over the years, we gradually extended the Fit Principle to people of all ages, and to relationship constellations such as that between a child, siblings and parents.

THE ELEMENTS OF THE FIT PRINCIPLE

Every person is unique.
Every person wants to satisfy their basic needs.
Every person wants to develop and use their competencies.
Every person develops their own ideas and convictions.
Every person has their own experiences, which help to shape their individuality.
Every person strives to exercise their individuality in harmony with their environment.

Humans can't live just any life; they can only live their own. Every person is unique in their essential make-up, and their path through life is just as unique as they are. The elements that make up their individuality have been presented in detail in Chapters 4–7. The illustration below shows how they interact.

The basic needs are the elements that determine a person's life to the greatest extent. Day after day, we all try to adapt to our environment in a way that allows us to satisfy our basic needs as well as possible. We use our competencies to do this, and are guided by our ideas. The extent to which we are successful in satisfying our needs depends on the environment: family, school or workplace, for example. And it also depends on the experiences we have had in the past, the situation in which we are currently living and the expectations we have of the future. Childhood poverty, or a guaranteed source of income, or the anticipation of a promotion and a salary increase have quite different effects on our life satisfaction. When someone manages to live in harmony with their environment, they feel mentally and physically well and have a good sense of self-worth and good self-efficacy.

In the sections that follow, we will look at what goes to make up a person's individuality, the areas in which the environment is particularly important, and – the core of the Fit Principle – how the individual and the environment interact.

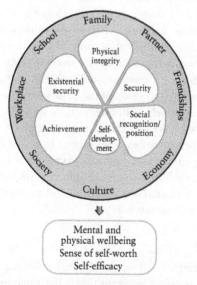

Figure 8.3: Fit constellation. Harmony between the individual and the environment. White: individual with different levels of basic needs; dark: environment.

The individual

Basic needs, competencies and ideas are the three fundamental areas that make up our individuality, and they have a substantial influence on how we are able to shape our lives. In addition to these areas, the experiences we have had over the years, and particularly during childhood, also influence how our individuality is configured.

We can measure the enormous significance the *basic needs* have for a person's life by the emotions that come with these needs: love and hate in relation to emotional security; greed and envy for existential security. Each of the basic needs is set at a different level for everyone (inter-individual variability). I have met some people who had an extreme need for emotional and existential security, and some whose emotional and existential independence astonished me. Some people devote their whole lives to achievement and throw all their resources into reaching a high professional goal. Others limit themselves to achieving just enough to make a living.

Basic needs are set at very different levels not just between people, but within the same individual. Every person has their own unique profile of basic needs, which they want to satisfy in their own way. And, depending on their profile, one person may find it easy to achieve harmony with their environment, while others find it much more difficult (for a detailed description of the basic needs, see Chapter 4).

Our *competencies* help us to satisfy our basic needs. Depending on how our basic needs are configured and the demands that life places on us, our competencies can be employed in very different ways. A secretary, for instance, uses linguistic competencies to write letters and documents; a politician wants to gain as much social recognition as possible and be elected to Parliament; and a writer's aim is to create a work that will outlast her.

The basic needs

- *Physical integrity.* We strive for physical wellbeing. We want to satisfy our elementary needs for things like food and sex.
- *Emotional security.* As children we are highly dependent on emotional security and affection. As adults we experience

emotional security as a feeling of trust, mutual support and accountability.

- *Social recognition.* We want to be respected for who we are and hold a social position that suits us. We look for a sense of belonging in our families and communities, in schools, the workplace and society.
- *Self-development.* In childhood, and to a lesser degree as adults, we strive to develop our competencies further and to work creatively.
- *Achievement.* We want to achieve the things we are capable of, based on our abilities, skills and knowledge.
- *Existential security.* We try to create the existential, material and social conditions in which we can satisfy our basic needs and keep hold of the things we have gained.

The competencies

- *Social competencies*, consisting of four components:
  - We use the elements of *nonverbal communication* to regulate our relationship behaviour. They are: facial expression, eye contact, voice, posture, gesture and distance.
  - *Caring behaviour* consists of the readiness to react to others' needs, particularly children and those who require help, and to support them in satisfying their needs.
  - *Imitative and social learning* is the innate ability to adopt other people's behaviour and internalize their values.
  - *Social cognition* helps us gain conceptual and emotional insights into our own state of mind (introspection) and that of others (extrospection). Empathy and symbolic ideas such as morality are important aspects of social cognition.
- *Language* is communication with a symbolic character. We have the unique ability to express symbolic ideas in formal structures (e.g. words and sentences). Language enables us to exchange information with other people on different topics and in different ways.
- *Musical competencies* consist of our sensitivity to rhythm, melody, harmony, dissonance and the timbre of notes. When we sing, dance or play an instrument, we use our musical competencies in combination with vocal skills and fine or gross motor skills.

- *Figural and spatial competencies* enable us to understand the appearances and spatial dimensions of the physical environment, and reproduce them creatively, for example by drawing.
- *Logical and mathematical competencies* in the broadest sense help us to gain insights into the properties of objects and abstract structures and how they interact. Understanding numbers (arithmetic) enables us to quantify sizes and amounts and use numbers operationally.
- *Temporal and planning competencies* involve our internal clocks, our ideas of time (for example, ideas about our own lifetime), and dealing with time in tasks such as planning work processes.
- *Motor and kinaesthetic competencies* include fine and gross motor abilities such as grasping and walking and skills such as writing. They also serve as an organ of perception, used when touching objects, for example (sensitivity to surfaces and depths).
- *Physical competencies* enable us to satisfy physical needs such as eating and drinking when we cook food, or to make an impression with our appearance using clothes and jewellery.

Like basic needs, competencies are set at very different levels from one person to another and within an individual. Everyone has their own profile of competencies, which essentially determines how they are able to adapt to and influence their environment. One particular challenge here is that – as students, for example – our task is not only to make the best use of our potential, but also to recognize and accept our limitations (for a detailed look at the competencies, see Chapter 5).

Ideas first develop from the specific experiences that children have with their social and physical environments. In the years that follow, they adopt an increasing number of ideas from their family, school and society. They internalize the attitudes to things like marriage or sexuality that are prevalent in their social environment. I only really became aware of the degree to which our ideas depend on our cultural surroundings, living conditions and the age in which we live in the early 1970s, when I travelled around Mexico. I spent time in some of the villages in the far south of the country, which were ruled by a matriarchy. The ideas about gender roles and bringing up children that held sway there were quite different from those in the rest of Mexico, where the power structure was patriarchal.

Different ideas and beliefs can also exist within a single culture. The differences can be traced back to the ideas that people adopted and internalized in their childhood. A child growing up in a strict Catholic family will develop a different morality to a child whose parents have a more liberal lifestyle. As adults our values may go on changing, depending on our relationship experiences.

Finally, ideas differ from one person to another because they are always also an expression of our basic needs and competencies. People with a great need for emotional security have different ideas about marriage and family from those who feel that need less acutely. People have their own ideas about what is required of them in school and at work, depending on the strength of their need for achievement. People who have a strong need for existential security place much greater importance on wealth and material security than people who feel less of a need in this area, though the latter group may have high expectations of the community they live in. A rich entrepreneur will have a different political position from a journalist with a very modest lifestyle who writes articles deploring the ever-widening gulf between rich and poor. And, not least, people have quite different ideas about the significance of the arts and sciences, depending on their own need for self-development and the levels of their competencies – figural and spatial or musical, for instance. We don't have a single idea (in religion, for example) that we can say is 'correct' and holds true for everyone on the planet. Our basic needs and competencies are too varied for that (for a detailed look at ideas, see Chapter 6).

People aren't just passive recipients of all that goes on in their environment. They seek out experiences that will help them satisfy their basic needs, develop their competencies and form their own ideas. One person might play golf, while another plays a musical instrument and a third does amateur dramatics. Over the years, the experiences people choose help to shape their individuality.

If we can gain a good understanding of the interaction between our basic needs, competencies and ideas on one side, and our experiences on the other, we won't just understand ourselves better: it will also help us to understand the thoughts and actions of the people around us.

Environment

Our basic needs and competencies evolved in an environment that was fundamentally different to the one we live in today. For 200,000 years our ancestors occupied modest areas of land and spent their whole lives in family clans and tribes. They were surrounded by the natural world. Chapters 1 and 7 describe the highly complex environment that humans have created in the space of a mere 200 years. Today, very few of us live in the natural world; nor do we live in communities of familiar people, but in an anonymous mass society. We no longer live in a single small area, but spend our days in a variety of spaces that are often separated by great distances, such as our family homes and workplaces.

These huge revolutions in society and the economy have made people permanently insecure. State and economic structures have become opaque to us. We are increasingly unable to keep up with the pace of scientific and technological progress, in IT for example, or the automation of work processes. A booming construction industry and our rapidly increasing mobility mean that our living spaces are constantly being reconfigured, and we no longer feel quite at home. But what worries people most is that the networks of long-term, stable and reliable relationships that existed in our original small communities have become much smaller and less reliable. In Chapter 10 we will take a close look at why today's society and economy are meeting the needs of humans less and less, and what we can do about it. But first we will explore ways to live a fitting life even in the highly complex environment we currently inhabit.

Exercising our individuality in harmony with the environment

Do people consciously plan their lives? I suspect they often believe so, but the reality is that they are permanently occupied with satisfying their basic needs – unconsciously, for the most part. Their individual profile of basic needs, along with their competencies and ideas, thus makes a crucial contribution to how people shape their lives. And the environment always plays a role in how successful they can be.

To understand the reciprocal relationship between individual and environment better, let's look at the world from the point of view of four people between the ages of thirty and forty, whom I got to know through the Zurich studies, my clinical work and my private life. Thanks to favourable living conditions, they have largely succeeded in exercising their individuality. These brief biographies describe four very different lives. Jakob trades in real estate, Hannes is a passionate sportsman, Erika is an artist and Sofia works in social care. A semi-quantitative evaluation has been used to record their basic needs and competencies. The scale on which their individual profiles are measured tells us how significant each of their basic needs is to them and how well-developed their competencies are. It ranges from 1 (very low) and 2 (low), through 3 (moderately low) and 4 (average) to 5 (moderately high), 6 (high) and 7 (very high). Chapter 9 gives a detailed description of the method used to evaluate basic needs, competencies and ideas.

Using the profiles of Jakob, Hannes, Erika and Sofia, we hope to answer the following questions: how much do their profiles of basic needs and competencies differ from each other? How were their ideas and beliefs formed? Can their biographies be explained by the levels of their basic needs and competencies and their ideas? To what extent have their basic needs, competencies and ideas determined their lives? What influence have their past experiences had on their basic needs, competencies and ideas?

Jakob, the real estate salesman

Jakob trades in real estate. He has his own company and is married with three children.

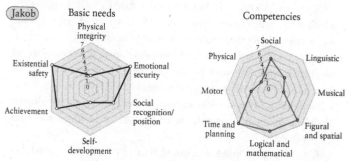

Scale: 1: very low, 4: average, 7: very high

Basic needs

Physical integrity: Jakob pays very little attention to his health. He is overweight and doesn't get enough exercise.

Emotional security: Jakob demands a lot of emotional security and affection from his wife, children and other relatives. He sees himself as a family man and a patriot.

Social recognition and social status: A strong position in the family and the company is very important to him. He avoids the public eye, as far as possible.

Self-development: his need for self-development is low. He doesn't have any hobbies, but is keen to keep improving his IT skills.

Achievement: Jakob is a workaholic. He often works evenings and weekends. Successful property transactions give him a deep sense of satisfaction. He expects his children to do well in school.

Existential security: Jakob works very hard. Providing material security for himself and his family is his real *raison d'être*, and it is what he puts all his time and energy into. He owns several properties and has amassed a substantial fortune.

Competencies

Social: Jakob is very caring towards his children. He takes a more authoritarian approach to his employees, but he is also good at responding to his customers' wishes.

Linguistic: Jakob is not very communicative and doesn't read much. His wife takes care of the company correspondence.

Musical: He has hardly any interest in music.

Figural and spatial: He has a well-developed visual imagination, which helps him evaluate properties. He has a good sense for the architectural styles that might be popular with his clients.

Logical and mathematical: A well-developed understanding of numbers helps Jakob with his bookkeeping. He is proud of having done most of the work to set up the company's IT system.

Temporal and planning: These competencies are very well-developed. They help Jakob draw up efficient work plans for building work and renovations.

Motor: His gross and fine motor skills aren't particularly good. He doesn't play any sports.

Physical: Jakob is tall and rather overweight, but his appearance is otherwise unremarkable.

Ideas

Jakob's approach to life is dominated by his desire for material security. When it comes to family, marriage and children, his attitude is patriarchal. He sees himself as the head of the family. He supports a Christian conservative political party, but isn't politically active. His *credo*: everyone is responsible for his own life.

Hannes, the sportsman

Hannes trained as an administrator. For the past ten years, he has worked as the HR manager in a large supermarket. But his main interest is more in sport than work. He lives with his partner and has no children.

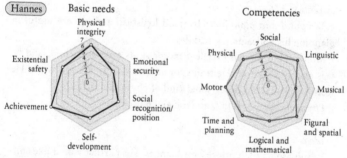

Scale: 1: very low, 4: average, 7: very high

Basic needs

Physical integrity: Physical fitness and health are very important to Hannes.

Emotional security: Hannes' need for emotional security and affection is about average.

Social recognition and social status: Hannes gets his recognition less from his work colleagues than from his fellow sportsmen, and he has a fitting social status among them.

Self-development: He makes a constant effort to improve the motor skills he uses in long-distance running and orienteering.

Achievement: Hannes takes pride in his work, but he has no desire to climb the career ladder. Sporting success is very important to him.

Existential security: Hannes has no material aspirations beyond a guaranteed income.

Competencies

Social: Hannes is quite skilled in his dealings with colleagues.

Linguistic: He can communicate well in German and Italian. He reads non-fiction books about nature and sport.

Musical: He occasionally listens to jazz music.

Figural and spatial: His spatial imagination is good, but not good enough to be really successful in orienteering, which he sometimes finds frustrating.

Logical and mathematical: These competencies are as well-developed as Hannes needs them to be for his work (buying and selling, for instance).

Temporal and planning: His good logistical abilities are useful in planning his colleagues' schedules.

Motor: Hannes had excellent motor skills even as a child, and he was also hyperactive. He trains for at least twenty hours a week. He feels happiest when doing some kind of physical activity.

Physical: Hannes is tall, athletic and muscular.

Ideas

Family isn't a high priority for Hannes. His partner would like children, but he doesn't really want them. He worries that being a father would leave him too little time for the sport he loves so much. He is very close to nature and supports ecological causes. He doesn't own a car and is a member of the Green Party. He is a devout agnostic.

Erika, the artist

Erika trained as an interior decorator and then took a course in graphic design. She would like to work exclusively as an artist. But she can't make a living from her art, so she also works part time as a

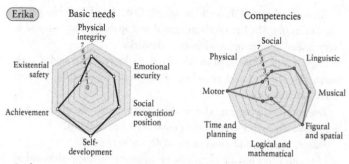

Scale: 1: very low, 4: average, 7: very high

graphic designer. Erika is single. She leads a very modest life. Her studio also serves as a flat.

Basic needs

Physical integrity: Erika is a vegan and looks after her health.

Emotional security: She is happy living alone. She has never had a long-term relationship. She has two cats.

Social recognition and social status: Erika receives a lot of recognition for her pictures, sculptures and installations. A certain status among artists is important to her.

Self-development: Her mission in life is to create art that will last, like Méret Oppenheim did. When she succeeds in this, she is really in her element.

Achievement: Before an exhibition, Erika works to the point of exhaustion. But she feels good doing it.

Existential security: Erika is always finding herself in dire financial straits, but that doesn't worry her.

Competencies

Social: Erika isn't very confident in dealing with other people; she avoids large events.

Linguistic: She enjoys conversations about things that interest her and reads plenty of literature.

Musical: Erika listens to a lot of music and loves to dance.

Figural and spatial: From a young age, Erika loved drawing, painting and crafts. She has excellent figural and spatial perception and is very skilled at turning her ideas into artworks.

Logical and mathematical: At school, Erika really struggled with maths. Her limited understanding of numbers keeps getting her into financial difficulties.

Temporal and planning: Her working methods are chaotic, which is a disadvantage when it comes to preparing for exhibitions.

Motor: Her fine and gross motor skills are very good.

Physical: Erika has an unremarkable appearance. She is slim and of average height.

Ideas

Erika is interested in esotericism and spirituality. She spent a few months in an ashram in India. She volunteers in an animal sanctuary. Erika doesn't regard herself as political.

Sofia, the carer

Sofia grew up in Africa and South America; her parents were missionaries. She spent many years working in the travel industry. Sofia is married to the mayor of a medium-sized city and has two children. She is active in institutions and charities that support disabled and socially disadvantaged people.

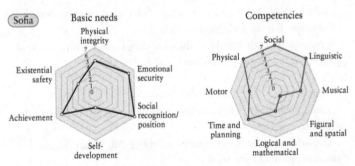

Scale: 1: very low, 4: average, 7: very high

Basic needs

Physical integrity: Nutrition and health are important to Sofia.

Emotional security: Family, relatives and friends mean a lot to her. She doesn't feel happy unless she has people she knows around her.

Social recognition and social status: Sofia places great importance on social recognition and the public's perception of her. She enjoys attending events at her husband's side.

Self-development: She has no real aspirations in this area and doesn't pursue any hobbies.

Achievement: Looking after her family's welfare and that of the disabled and socially disadvantaged people she cares for fills her with a great sense of satisfaction.

Existential security: Sofia doesn't worry about this. She relies entirely on her husband for her existential security.

Competencies

Social: Sofia is a real 'people person', and her behaviour is extremely caring.

Linguistic: She speaks four languages almost perfectly.

Musical: Sofia sings in a church choir, less for the music than for the sense of community.

Figural and spatial: Her orientation skills are very limited. She relies on a satnav when driving.

Logical and mathematical: Sofia is not very competent in this area.

Temporal and planning: She is good at planning and organizing events.

Motor: Sofia's motor competencies are average.

Physical: She is attractive and approachable. She always looks very well-groomed.

Ideas

Sofia campaigns for social justice in the world and dialogue between religions. She is active in the Social Democratic Party and supports her husband at political events.

Jakob, Hannes, Erika and Sofia have very different profiles of basic needs and competencies. To make the differences clear and enable us to compare them more easily, the figures are presented in a table below.

Levels of basic needs and competencies for Jakob, Hannes, Erika and Sofia. Scale: 1: very low, 4: average, 7: very high

	Jakob	Hannes	Erika	Sofia
Basic needs				
Physical integrity	2	6	5	5
Emotional security	7	4	4	6
Social recognition and status	4	5	5	7
Self-development	2	5	7	2
Achievement	6	7	6	6
Existential security	7	5	2	3
Competencies				
Social	5	5	3	7
Linguistic	3	6	5	7
Musical	2	4	6	4
Figural and spatial	6	6	7	1
Logical and mathematical	6	5	1	3
Temporal and planning	7	6	2	6
Motor	3	7	7	4
Physical	2	6	2	7

Physical integrity (fitness, for instance), is a very low priority for Jakob (2). Hannes (6) sees it as very important, and for Erika (5) and Sofia (5) its significance is about average. Jakob (7) and Sofia (6) are very reliant on emotional security, while Hannes (4) and Erika (4) need much less of it. Jakob (4) has an average need for recognition, though his clients' opinion is very important to him. Sofia (7) lives for the social recognition she receives from the public and through her activities in institutions and charities for disabled and socially disadvantaged people. The recognition that Hannes (5) and Erika (5) crave is more specific, for their sporting and artistic achievements respectively. Jakob (2) and Sofia (2) don't feel much of a need to be creative, but this need is very high for visual artist Erika (7) and slightly above average for Hannes (5). All four of them want to achieve (6 or 7), though each in their own way. Jakob's sense of achievement comes from accumulating material wealth, Hannes' from doing well in sports, Erika's from her paintings and sculptures and Sofia's from her work in the social-care sector. Jakob (7) feels that providing existential security for himself and his family is his real *raison d'être*, and he works hard on it. Hannes' concern for existential security is slightly above average (5), and Sofia's is slightly below (3). For Erika (2) it holds little or no importance, although she is always finding herself in financial difficulties.

The competencies of the four are also set at very different levels. Jakob's social and linguistic competencies are average (5) and below average (3) respectively. His writing skills are quite weak, so his wife takes care of the company correspondence for him. Sofia's social and linguistic competencies are extremely well-developed (7; 7), while her figural and spatial competencies are very weak (1). She has no sense of direction. The other three have good or very good figural and spatial competencies, though they use them in different ways. Jakob's (6) help him to value properties; Hannes (6) uses his for orienteering, and Erika (7) creates sculptures and installations. Jakob has good logical and mathematical competencies (6) and very good temporal and planning competencies (7), which are extremely useful in creating workflow plans and solving IT problems. Erika (1; 2) has significant weaknesses in this area. Having to prepare work for exhibitions is a real challenge for her. Jakob's motor competencies are not well developed (3), while Hannes' and Erika's (7) are excellent. Hannes uses his to run races, and Erika's help her to dance and make art. And, finally, Jakob (2) and Erika (2) have very low physical competencies, while Sofia's are very highly developed. When she makes public appearances, Sofia can rely on her attractive appearance and her approachable manner to win people over.

Basic needs and competencies are set at different levels not only from one individual to the next (inter-individual variability), but also within individuals (intra-individual variability). Jakob's basic needs for emotional security (7) and existential security (7) are very pronounced, while his need for physical integrity (2) and self-development

(2) are low. Erika and Sofia's profiles are also irregular, but Hannes' basic needs and competencies are all developed to a fairly equal extent.

What about their ideas and convictions? These also vary greatly between the four. Jakob has a conservative attitude to family and society. He places great value on clear social hierarchies and a free market economy. He is a practising Catholic with a traditional Christian approach to life. Hannes is an agnostic who cares about the natural world and is a member of the Green Party. Erika is apolitical and interested in spiritual ideas. Sofia's life revolves around social issues. She is involved in creating dialogue between religions.

How did Jakob, Hannes, Erika and Sofia arrive at their ideas and convictions? Like all children, they took on ideas from their social environment. Sofia internalized her parents' social and religious beliefs. Over the course of our four subjects' lives, their ideas developed based on their basic needs and competencies, and the experiences they had. This process of adaptation can mean that siblings and even twins develop in increasingly different directions, and grow further apart over the years (see Chapter 3). Jakob and his brother, Martin, have different opinions when it comes to politics. Jakob's approach to life is conservative in every respect, while his brother and his parents are much more socially minded. As a teacher, Martin does a lot to help the children of migrant families integrate in schools. It doesn't necessarily follow that rich people like Jakob will always embrace a conservative attitude. Some very wealthy people have high social competencies, and take a liberal or even a socially progressive approach to life. On the other hand, not all teachers are as liberal and socially minded as Martin. Some even take an extremely conservative position on social and education policy issues.

At various points our basic needs, competencies and ideas will make us rethink our lives and perhaps try taking a different path. Jakob's parents thought it only natural that he would become a teacher, like them. They persuaded both him and his brother to train as teachers after leaving school. But, having completed several years of training, as soon as Jakob found himself in a classroom he quickly realized that teaching was too much of a challenge for his limited social and linguistic competencies. The hours he had to spend in front of a class of children, and the constant effort to communicate with them and hold their attention, left him exhausted within a year. He left to work in a bank, and a few years later set up his own property business. Hannes grew up in the countryside. His parents were farmers and ran an agricultural business. They expected Hannes to take it over when he grew up, but farm work didn't suit him. He was glad when his younger brother expressed an interest in taking over the farm, and Hannes was able to do a business apprenticeship. But Hannes' love of nature never left him. Erika and Sofia have mostly been able to lead lives that fitted their individual genotypes. Erika's parents managed a care home for disabled children. They never expected Erika to follow in their footsteps. They recognized their daughter's creative talent early on and supported her when she trained as a graphic designer and artist. At school, Erika struggled with maths. Her parents and teachers showed great understanding for her specific learning difficulty, and with their support Erika learned to cope with it. Sofia spent her childhood in Africa and South America. Her parents were missionaries and worked in development aid. The wealth of different cultural experiences Sofia gained while living with them suited her basic needs and competencies very well. She loves interacting with people from different cultures and speaks four languages perfectly.

We often describe people as having certain personality traits. We might call them gregarious or shy, tidy or chaotic, creative or unimaginative. But these aren't independent characteristics: they are the expression of that person's profile of basic needs, competencies, ideas and emotions. The overall impression we receive of someone comes less from a single basic need, a specific competency or a particular idea they hold – though artists and scientists, for example, may be exceptions to this rule. Our impression of someone comes from how their individual profile is put together. Jakob gives the impression of being a successful real estate trader because he is driven by his great need for existential security and has highly developed figural, spatial, logical, mathematical, temporal and planning competencies. Sofia's personality is shaped by her strong need for social recognition, her well-developed social and linguistic competencies, and her attractive appearance. People like Hannes, with his regular profile, give the impression of being stable characters. Erika, on the other hand, seems unstable. With her one-sided profile and a boundless desire for self-development, she runs the risk of ending up in difficult situations. If she can't make a success of her art, then adapting to her environment will be much more difficult for her than for Hannes.

Is it possible to live a life that goes against our basic needs and competencies? Jakob, Hannes, Erika and Sofia all have profiles of basic needs and competencies that fit their biographies well. It's difficult to imagine Hannes leading Jakob's life, and vice-versa. The same applies to Erika and Sofia. If they had been forced to, all four of them would have been unhappy. Some people try to emulate a great role model, such as a talented pop star or an innovative, wealthy IT entrepreneur, even though they don't have anything like the right set of basic needs and competencies. They quickly come up against their limitations, which they cannot overcome no matter how hard they try. We all have to accept that our basic needs and competencies are limited. We can only lead our own lives. Being true to ourselves is a fundamental prerequisite for leading a fitting life.

FREE WILL AND
THE MEANING OF LIFE

Every person has free will

The only person who is himself free is that person who wishes to liberate everyone around him.

Johann Gottlieb Fichte (1762–1814)[12]

Some readers may have arrived at the oppressive conclusion that our thoughts and actions are largely predetermined by our basic needs, competencies and ideas – and that the idea of free will is mere wishful thinking. It is certainly true that we don't have the kind of agency that allows us to make completely free decisions, independent of our needs, competencies and ideas (see Chapter 6).

But the Fit Principle works on the basis that every person does have free will, in the sense that they want to satisfy their basic needs, use their competencies and live according to their ideas. If they are denied these things, they feel their free will has been restricted. Even small children show great strength of will when they're prevented from having their own experiences, such as crawling up and down the stairs. They protest by shouting and screaming, and try with all their might to assert their will. Older children can also muster an impressive level of determination, in learning to ride a bike, for instance. And some adults have extraordinary willpower, too – like the young entrepreneur who builds up his own IT company.

People aren't uniformly strong or weak-willed. Their willpower varies according to the area of basic needs, competencies or ideas in which they want to assert themselves. Jakob, as we have seen, has very pronounced needs for existential security and achievement. He has tremendous drive when it comes to his company, but lacks the will to apply his motor competencies to any kind of sporting activity. Hannes' motor competencies are very well developed. His will to achieve great things in sport is very strong, much stronger than his will to climb the career ladder. Jakob and Hannes don't just have different religious and political ideas: their willingness to stand up for their convictions is also different. Jakob is a passive member of a Christian political party, while Hannes campaigns actively for ecological causes. A person's will is as varied as their basic needs, competencies and ideas, and always depends on the situation in which they find themselves.

If our will is restricted, we feel we have been called into question and that other people have control over our thoughts and actions. The chance to live a self-determined life is a precious commodity, which we must not only make use of ourselves, but grant to everyone else as well, by respecting their will. We must stop asserting our own will at the point where it begins to restrict other people's freedom. We only feel free – as the philosopher Johann Gottlieb Fichte notes so perceptively – when we allow our fellow humans their free will, too.

The search for the meaning of life

Becoming who we are is our life's goal and our life's meaning in one.

Helga Schäferling

We have no way of knowing, but it's highly unlikely that animals – even our closest relatives, the primates – think about the meaning of life. That requires consciousness and a set of sophisticated ideas about things like birth, life and death. Even human children lack these ideas in their first years of life. Then they begin to ask 'why' questions, such as why the apple tree in the garden sheds its leaves – and the questions never stop. But it's only in adolescence that they start asking about the meaning of life. And, at that point, many young people experience a kind of identity crisis. What do I want to do with my life? What is my *raison d'être*? What is worth fighting for?

The meaning of life takes a central place in philosophy and religion. Over the past 2,000 years, philosophers have come up with various answers to the question of what constitutes the meaning and purpose of human life. For the philosophers of antiquity, the primary purpose of life was to achieve true happiness. To reach this goal, they urged people to take meaningful actions, and encouraged them to practise virtues such as wisdom, justice and moderation, and shun vices like greed, lust and power. For religions, the meaning of life was provided by their spiritual truths – in Christianity, for example, meaning is found in the search for enlightenment and the union with God. The Catholic Church still claims to be infallible when it comes to lending meaning to human existence.

Most philosophical schools and religions have a practical as well as a spiritual level on which they address people's basic needs. The hedonists, for instance, emphasize seeking out and experiencing sensual pleasures. In Christianity, the longing for emotional security, care and affection is expressed in the love of God and man. Ever since ancient times, philosophers have been speaking up for justice as the only way to guarantee the dignity of the individual and a peaceful co-existence with our fellow man. Another objective for many philosophers is to discover our true selves and develop our talents. For Friedrich Nietzsche, the meaning of life lay in man developing into a higher life form: the superman. With its character-building potential and creative power, leisure was also cited as a source of meaning by philosophers from the ancient Greeks onwards. During the Reformation leisure fell into disrepute, and from then on people's lives were shaped by the desire to achieve. In Protestantism (and Calvinism in particular), work is God's purpose for us in life. The morality of work that goes with this idea is based on the virtues of diligence, self-discipline and frugality. According to the sociologist Max Weber, this morality made a fundamental contribution to the Industrial Revolution and modern capitalism. As we strive to achieve existential security, philosophers and the founders of religions warn us of the darker sides to our quest: acquisitiveness and wealth, exploitation and poverty. The significance of philosophy and religion lies not only in the rational insights and spiritual messages they provide, but also in the practical way they help us to satisfy our basic needs.

There is a telling phenomenon that can be observed both in individuals and in a collective. As long as life is good and people are mostly managing to live in a Fit constellation, they give little thought to the meaning of life. But if things are going badly, if an individual falls seriously ill or a population is suffering from extreme poverty and oppression, they begin to search for meaning. The sick ask themselves whether they have led meaningful lives. The population search for something that will lift them out of their misery, and find an answer in communism, for example, as an antidote to exploitative capitalism. From the Fit Principle point of view, the meaning of life lies in creating an image of humans and the world that suits you as an individual, and above all in the never-ending effort to lead a fitting life.

9

Misfit Constellations

The Fit Principle is all about tackling the Misfit situation by questioning your current way of life.

If I keep from meddling with people, they take care of themselves,
If I keep from commanding people, they behave themselves,
If I keep from preaching at people, they improve themselves,
If I keep from imposing on people, they become themselves.

Lao Tzu (c. 6th or 4th century BC)[1]

No one, I suspect, has ever succeeded in living in a permanent Fit, in constant harmony with themselves and their environment. Misfit situations that individuals manage to deal with themselves are part of everyday life. They are an incentive to pay more attention to a particular area of life. This process comes with 'eustress' (from the Greek *eu*, meaning good, and stress, reaction to pressure), which makes us more focused and capable in body and mind. Misfit situations force us to keep checking the validity of our habitual behaviours, beliefs and goals, and adapt to changing circumstances. We rise to challenges we can overcome and are pleased when we succeed. We don't always have to be successful to stay motivated, but we need to succeed often enough not to be discouraged. Eustress doesn't wear us out, and has no long-term negative consequences. It doesn't affect our mental and physical wellbeing, or our sense of self-worth and self-efficacy. Misfit situations that cause only eustress are essential if we are going to make any lasting changes to our lives. Our evolutionary experience makes humans capable of solving most, if not all, Misfit situations on their own and – as Lao Tzu remarked 2,400 years ago – of becoming more ourselves each time.

In Chapter 9 we will look at Misfit situations that go beyond eustress and trigger *distress*, affecting our physical and mental wellbeing so badly that we become incapable of thinking or acting. In so doing, we will cover the following questions: how do people react to excessive stress? How do Misfit constellations arise, and why do some people find Misfit situations harder than others? What effect does being in a Misfit situation have on people? And how can a Misfit situation be resolved?

MISFIT CONSTELLATIONS ARE AS VARIED AS HUMANS THEMSELVES

From eustress to distress

Our lives are shaped by challenges both large and small. We have periods of being under-challenged, when we are doing things like household chores. And when more is demanded of us – in the workplace, for instance – then our eustress increases and we achieve more. But if these challenges go beyond a certain limit, which some people reach sooner than others, then eustress tips over into distress. If these challenges keep coming, or become even more difficult, our ability to achieve drops sharply and we feel overwhelmed. There comes a point where we feel exhausted and can achieve nothing at all. A Fit has become a serious Misfit situation, in which there is little harmony between a person and their environment. The Misfit can be caused by the individual or the environment, but most often the causes are to be found in both.

We are most familiar with Misfit constellations as a feeling of being overwhelmed, which can happen when we're trying to achieve things – at work, for example. But each of our basic needs can be affected by a Misfit. If children are emotionally neglected, it will have a negative effect on their wellbeing. Adults experience stress when

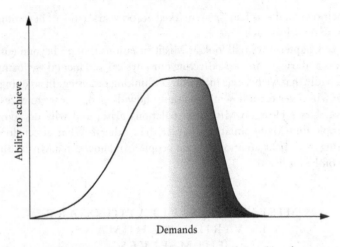

Figure 9.1: Fit and Misfit. Level of achievement and transition from eustress (light) to distress (dark) as demands increase.

they aren't getting any recognition in the family or the workplace, when they can't make proper use of their abilities, or when they face existential insecurity because their wages might not see them through until the end of the month.

When people first find themselves in a Misfit situation, they try to live with the distress and ride it out. They cling to the hope that things will get better. After a divorce, a mother might try to manage her job and care for her two children alone. She will feel overwhelmed, but she will also hope that in time this double burden will ease. Most people react not only by riding things out, but also by working harder and trying to compensate for the problem. If a relationship is threatening to fall apart, one partner will start paying more attention to the other and trying to please them in any way they can. Employees will start working evenings and weekends to keep their employer happy and avoid being fired. If someone loses social recognition in the workplace, they will seek more attention from their family and friends.

But as stress increases, so does the impact it has on people's mental and physical wellbeing. Some manage to recover by using relaxation techniques and working on their mental and emotional strength, perhaps by taking up one of the many forms of yoga and meditation. Others opt for a physical treatment such as craniosacral or shiatsu therapy. And some people attend courses and seminars (increasingly with an esoteric and pseudo-scientific bent) that promise to teach coping strategies and ways of becoming more stress-resistant. In recent decades, our tremendous need to recuperate from these Misfit situations and develop coping strategies has given rise to a booming coaching and wellness industry. And many of these businesses don't just promise to restore mental and physical wellbeing: they aim to release additional strength to help people overcome future challenges and achieve even more. But if someone still feels overwhelmed despite their best efforts, or if they face even greater demands and ever-increasing pressure, eventually their performance will suffer. The Misfit is manifested both mentally and physically.

How can we tell when we or people we know are in a Misfit situation? How does distress make itself known? First our wellbeing, our sense of self-worth and self-efficacy are impaired. And if the Misfit gets worse, people may display the following symptoms, depending on their mental and physical disposition:

A change in *temperament* – becoming more impulsive and less controlled, for instance.

Emotional disorders, which take the form of anxieties. People become more moody and irritable.

Social competencies such as empathy and caring behaviour are reduced. People are concerned with themselves first and foremost.

Decreased *willingness to learn and achieve*.

Increased *unusual behaviour*, which may include aggression or social withdrawal.

Psychosomatic disorders, including sleep disorders or digestive problems.

Addictive behaviour, such as excessive alcohol or drug consumption.

Ultimately, *states of exhaustion* such as depression or burnout. People are unable to cope with the most basic everyday challenges. Emotionally, they feel completely drained.

The way in which people react to Misfit situations depends on their individual mental and physical disposition. They can become angry and irritable, or withdrawn and suffer their stress in silence, or they

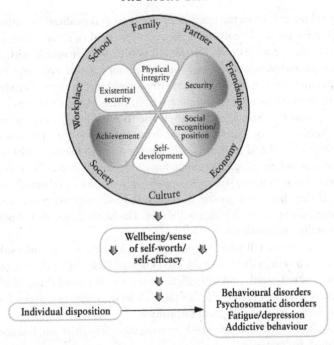

Figure 9.2: Misfit constellation. Individual: the size of the basic needs represents how strongly they are felt; white: proportion of basic needs satisfied; dark: proportion not satisfied. In this example existential security, physical integrity and self-development are sufficiently satisfied; emotional security, social recognition and achievement are only partly satisfied. Grey: environment.

might get upset and anxious. Some people become less willing to learn and achieve. Most people also experience psychosomatic symptoms such as migraines and sleep disorders. Many resort to medication – initially, over-the-counter remedies. If that doesn't help, some will go to the doctor and get a prescription for physical symptoms like headaches and heartburn, sleeping pills to help with insomnia, or antidepressants and stimulants to combat depressive moods and list-lessness. A considerable percentage of people suffering from stress tend towards excessive alcohol consumption. The use of performance-enhancing drugs (neuro-enhancers) such as Ritalin and sedatives such as Valium is widespread among overworked managers and students worrying about exams. If a serious Misfit constellation lasts for an extended period, it can lead to a state of general mental and physical exhaustion, which the medical profession calls 'reactive depression' or 'burnout'. People affected by burnout are literally brought to a standstill, and require extensive psychotherapy and often an extended stay in a psychiatric hospital. At this point, if not before, there really is no option but to fundamentally rethink your life.

How children deal with Misfit situations

Children can only satisfy their own basic needs in a very limited way, and in the first years of life they can't satisfy them at all. They are reliant on other people, mainly parents and attachment figures such as nursery workers. And so when children find themselves in a Misfit situation, it's almost always the environment that is at fault, and not them.

Misfit situations can have a negative effect on each of the six basic needs, impairing children's wellbeing and causing symptoms of mental and physical distress. A toddler might be receiving too little emotional

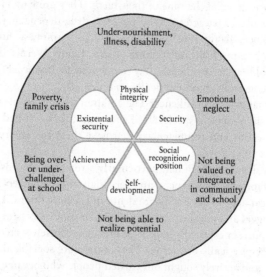

Figure 9.3: Risk factors in childhood that can trigger a Misfit situation.

security and affection from his family or nursery. He becomes particularly clingy and refuses to eat. A child's physical wellbeing can be so badly affected by his asthma that his motivation to learn and achieve decreases. Another child might receive too little recognition from her friends because a motor disability means her participation in group activities is very limited. A child's self-development can be affected if her musical talent is not acknowledged and encouraged. Even in prosperous countries, the wellbeing and development of millions of children is hampered because their families live in poverty and are not sufficiently integrated into society.

Experts have spent decades puzzling over why some children's mental and physical wellbeing is severely impaired by growing up in difficult circumstances, while other children living in comparable conditions are affected much less or not at all. The developmental psychologist Emmy Werner found an explanation for this phenomenon, which she laid out in *The Children of Kauai* and other books. She conducted a comprehensive study that tracked the development of nearly 700 children born in the same year from 1955 to 1995.[2] The children lived on Kauai, an island in the Hawaiian archipelago that was very isolated at the time of their birth. They grew up in difficult circumstances, characterized by poverty, drug-dependency, neglect and violence. To the research team's surprise, around a third of the children became fully functioning adults. Emmy Werner described these children as 'resilient': they were able to bounce back from adverse conditions both physically and mentally.

But what was this resilience? The children who overcame their difficulties didn't have a special 'resilience gene'; they just had a profile of basic needs and competencies that gave them the ability to cope with life.

The resilient children had particularly strong basic needs, such as the need for self-development, and very high-level competencies. They had a great need to develop their talents and achieve things. They were good singers and musicians, for example, or very successful in a particular sport, or especially good at expressing themselves creatively, or they were physically attractive. They were more sociable than other children and actively sought out trusted people who became attached to them and supported them in their development. These characteristics made the children appealing to people outside their families, including music teachers and sports coaches. Adults who weren't their parents then became important attachment figures, and these adults enjoyed encouraging the children's development and watching them grow up to fulfil their potential. This broad foundation of support helped them achieve good exam results, find an apprenticeship and embark on promising careers.[3] The support they received also came with a better level of social integration, which lowered these children's susceptibility to criminal activities, drug dependency and mental illness.

What makes resilient children so strong

- Resilient children tend to be more intelligent and have higher competencies than less resilient children.
- They are more interested in people, things and ideas. They are eager to learn, and most of them enjoy going to school.
- Resilient children often have higher than average social competencies. They pay attention to other people and react positively to the attention they receive.
- They are more socially adaptable and 'biddable'. They try to fulfil adults' expectations of them.
- They are interested in attachment figures outside the family. They actively seek out relationships with adult attachment figures who will take an interest in them and give them support.
- Contrary to a widely held prejudice, they are not 'tough'. They are actually more willing to ask for help and admit to weaknesses than other children are.
- Resilient children display a high level of impulse control and are prepared to defer rewards.
- They have a realistic self-image and firm ideas about the future.
- When they leave school, they often also leave an unpleasant family situation and find an environment that corresponds more to their needs.
- As adults, they have a strong sense of self-worth and greater self-efficacy.

But resilient children were only successful in their efforts to form relationships and find support if there were people in their social environment willing to take them on. In the Kauai study, and in other studies, this kind of environment had the following features:

The families and communities where resilient children grow up

- Parents of resilient children are often more educated than parents of less resilient children.
- They are friendly, sympathetic and supportive, and take an interest in their children's lives.
- They themselves are also resilient to some degree and set an example of how to cope under difficult circumstances.
- The social cohesion created in organizations such as schools and sports clubs provides children with emotional security and social recognition, in addition to the developmental experiences they have there.
- Social cohesion requires children to display empathetic behaviour and be willing to take on responsibility.
- Trusted people give the children attention, serve as role models for coping with life, and help them to have the learning experiences they need. They spark an interest in reading, for instance, by giving children access to a library.

The parents of resilient children place great value on a good education, even if they themselves haven't had the opportunity to attend a good school, or to go to school at all. They hope their children will do better than them. Within the family, older siblings, grandparents and relatives also play an important role. Fewer resilient children than less resilient children grow up in single-parent families. In reliable, tight-knit communities teenage girls are less likely to get pregnant and boys are less likely to turn to crime or drugs. Parents of resilient children frequently engage with church groups, schools and other community activities that provide children with a sense of emotional security and shared values.

In the Kauai study and other similar investigations, researchers came to the unanimous conclusion that, when children thrived under difficult circumstances, it was partly due to the children themselves, with their advantageous profiles of basic needs and competencies, and partly because they had supportive people around them. The interaction of the children and their environment produced an upward spiral, which gave their development an additional boost. This is not just the case for children who grow up on a poor island such as Kauai, but for those who live in affluent societies as well. And not just the talented ones, either, but every single child.

Most children have average levels of basic needs and competencies. They don't have the confidence to approach attachment figures and fight to satisfy their basic needs. If the environment doesn't provide for them adequately, they can enter a downward spiral. These are the children we must take special care of. We can't wait for them to come to us with their needs. We have to pay more attention to them and give them all-round support. This can involve more effort on our part, but it can also fill us with a deep sense of satisfaction, because these children are especially grateful for a trusting relationship and the right kind of support.

In our achievement-focused society, many children whose abilities are merely average find themselves in Misfit situations, because they are unable to satisfy the expectations of their parents, teachers or even society as a whole. Much as these children would like to achieve more, they don't have the competencies to do so. Even those who live in privileged circumstances (supportive parents; a good education system) aren't able to develop above and beyond their potential (see Chapters 2 and 3) – and some parents and teachers refuse to accept this fact. Children become insecure and anxious when they sense that they can't fulfil their parents' and teachers' expectations. And this state of mind can make them unable to achieve even the things they really should be able to manage. The pressure on children from family and society has become so great that it is having a lasting effect on some children's development.

Finally there are very vulnerable children, whose limited abilities mean they are at risk of falling into Misfit situations over and over again. Karim was one such child who was brought to my consulting room. He was of average intelligence, but suffered from a slight motor disability. During his first years at school he had become increasingly overstretched by writing and drawing tasks. Karim's disability didn't just affect his physical integrity: it was also hampering his ability to achieve. He was finding it harder and harder to keep up with the other children, which was making him feel discouraged. His insecurity had developed into a mental block on learning in other subjects as well. His wellbeing, sense of self-worth and self-efficacy were affected, and he was suffering from a severe sleep disorder. Karim's parents and teacher weren't prepared to accept his physical disability: they wanted us to treat and cure it. I told them that the disability couldn't be eliminated. The goal should be for an occupational therapist to help Karim learn to deal with it as best he could and to accept it. His parents and teachers could help by having patience with his slow pace of work and praising his efforts rather than his achievements. In this way, Karim would grow into a self-confident adult.

All children, whether they are gifted or averagely well equipped or have physical or learning disabilities, want their basic needs to be satisfied. They all want to be able to develop and use their competencies. We can support them most effectively by accepting their uniqueness and arranging their environment in a way that allows them to develop on their own terms.

Sigmund Freud taught us that traumatic experiences in childhood can have long-term effects on a person's mental wellbeing. Sexual abuse in childhood can, in the worst case, affect someone's emotional stability and capacity to form relationships for the rest of their lives. Misfit situations don't just affect children's emotional and social needs: they can affect every one of their basic needs. The effects of mental and physical cruelty in the family can last for years, but so can debilitating experiences such as being overtaxed by schoolwork, being ostracized by teachers and peers, receiving no encouragement to develop your talents, or living in poverty and internalizing your parents' existential anxieties. Supporting children's development so that they find themselves in as few Misfit situations as possible, and helping them find their way out when these situations do occur, isn't just important in childhood: the effects of that support last much longer.

How adults get into Misfit situations

The same is true of adults as it is of children: Misfit situations can affect people differently depending on their profile of basic needs, competencies and ideas, their previous experience of Misfit situations and the burdens they are currently carrying. Being made redundant can make one person feel existentially insecure, while someone else might miss the satisfaction of a job well done. A third person might worry about the loss of professional status, and a fourth might be affected in all of these areas.

And, like children, adults have different levels of resilience. Some refugees with highly developed competencies arrive in Europe and, with the right support, are very quickly integrated into society and begin working their way up into senior positions. But there are far greater numbers of refugees with average competencies, and others who have some kind of disability. They are all especially reliant on our support.

These days the most common problem people face isn't making something of themselves under poor living conditions, but surviving in a very competitive society. How successful people are depends on their profile of basic needs and competencies. People with very strong basic needs and above-average competencies are often convinced that success and the avoidance of major Misfit situations are their just deserts. This belief helps them claim a high social status, an exorbitant income and all kinds of privileges. But people with average basic needs and competencies make just as much effort, if not more, than those whose genotype and living conditions have put them ahead of the game. And then there are those people who find it particularly difficult to make something of themselves in this society due to a physical or learning disability. Each of the basic needs and competencies can be affected by a Misfit situation. Physical integrity can be diminished by a condition like diabetes or rheumatism, or by life taking an unhealthy turn – into alcoholism, for example. This can lead to a Misfit situation in the family and at work. An averagely competent employee might be overtaxed by what a manager regards as a very reasonable workload. The overload is particularly damaging if the employee and the manager react to it with criticism and rebuttal rather than understanding. In the same situation one person might feel happy and another might suffer, depending on how their basic needs and competencies are configured. Take a nursing home, for example, where one elderly person is still surprisingly sociable and communicative with other residents and staff, and receives a lot of

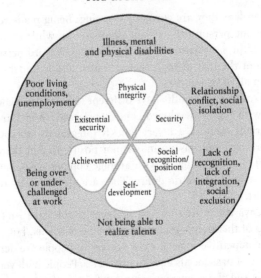

Figure 9.4: Risk factors that can lead to Misfit situations in adulthood.

attention as a result; the person in the next room might have a more limited ability to form relationships, and therefore receive too little care and attention. Some people are prevented from using their competencies as they would like by social and economic conditions. Immigrants often find themselves in Misfit situations like this. A neighbour of mine who is originally from Poland told me why she took a job in Switzerland for which she is over-qualified. She had studied psychology in Poland, but couldn't find a job in that field. In Switzerland she works as a housekeeper for an elderly woman in order to fund university places for her two sons. I was driven to an event in Berlin by a taxi driver who had been a doctor in Pakistan. He was forced to flee the country for political reasons and now, since he can't work as a doctor in Germany, he has to support his family by driving taxis.

Misfit situations can also arise when people are guided by ideas and examples that don't fit with their individual basic needs and competencies. Parental values, such as marriage being for life and divorce being unthinkable, can make a daughter with an unsuccessful marriage behind her feel like a failure and suffer terrible guilt. A young man might study architecture because his father is a top architect and he doesn't want to disappoint him, even though he doesn't have the necessary competencies for it. Misfit constellations can also arise inadvertently; for example, when a maths teacher's skill and dedication convince a student that he should apply for a maths degree, although he doesn't have the talent to complete it.

Vulnerability is more widespread among adults than is generally assumed. If we work on the principle that, for each of the eight competencies, a minimum of 3–5 per cent of people have a weakness in that area, then more than a third of all people are affected by a specific learning difficulty (see Chapters 2 and 5). These limitations affect people to varying degrees, depending on the competency involved. Weaknesses in reading and writing have a greater impact than low or no musical competencies do. In our society, linguistic competencies are a far better guarantee of existential security than many others – even if you happen to have a great musical talent. Often the person affected and the people around them aren't even conscious of their limitation. A worker in a logistics company might be excluded by his colleagues because his limited planning and organizational abilities don't allow him to fully integrate into the work processes. He feels like a failure, and his colleagues regard him as a slob. Competencies are set at different levels within individuals, so it might be that one competency is very well developed, while another is almost non-existent. A mayoral candidate may win over voters with his engaging appearance and his eloquent speeches – but, once in power, he will disappoint citizens with his poor social and planning competencies.

Misfit constellations become more acute when more than one basic need is affected; for example, if an adult is suffering from a chronic illness and then becomes unemployed and socially isolated. Misfit constellations have particularly serious consequences for asylum seekers and refugees, because all their basic needs are more or less affected: a high level of existential insecurity caused by the loss of income and property; a diminished state of health caused by hunger and illness; the loss of emotional security from a family becoming separated and relatives lost; a lack of social and professional integration and access to the education system. People in this situation feel helpless and as though they have lost control of their lives, which can have a disastrous effect on their wellbeing. These people rely on us not only welcoming them, but providing them with all-round support.

We all have our strengths and weaknesses. We draw on our strengths, and we are never rid of our weaknesses. We just have to learn to live with them, though this task is not made easy for us. Society and industry are only interested in our strengths, meaning that increasing numbers of people feel distressed by their weaknesses. The world of business in particular shows very little understanding for these people – although they depend on consideration and concessions in order to manage their lives successfully. They don't want to be patronized; they just want to be able to live a self-determined life. Misfit situations are becoming increasingly frequent in our society because of the decline in communities where everyone knows each other. It is only within these communities that people are prepared to accept the weaknesses of their fellow humans and help them when they're in trouble. The less people are able to satisfy their basic needs and develop their competencies, the more important it is that they have the support of a community. We will take a closer look at how living with familiar people can help people to achieve harmony with their environment in Chapter 10.

MANAGING MISFIT SITUATIONS

We make the changes that small, everyday Misfit situations require of us without even thinking about it. We are guided by our gut instincts. We sense intuitively what is going wrong, and try to correct it. If we neglect our emails for too long, we feel a vague sense of unease. Finally, we sit down at the desk and get to work. But if warning signs like headaches, insomnia or even depression start to appear, we know we have got into a more serious Misfit situation to which we must pay some proper attention.

Understanding Misfit situations

The first step to finding your way out of a Misfit situation is to ask yourself how it came about; how your own basic needs, competencies and ideas are configured, and why they don't fit with your environment.

Tanja's life story shows us how a Misfit situation can arise, and how it can be resolved.

Tanja is twenty-two years old, and is currently being treated for anorexia as an outpatient at a psychiatric clinic. She tells the doctor that her parents divorced when she was three and her brother, Aldo, was six. She only saw her father a few times over the years that followed, though he used to send her little presents for her birthday. Her mother was an administrator for a large business. She provided for herself and her children, as their father's alimony payments were irregular. They had to scrimp and save; family holidays were a rarity. Tanja's mother was chronically overstretched by the demands of work and family. In the evenings and at weekends she was exhausted and found it difficult to take care of the children. When she was at work they were looked after by their maternal grandparents and a series of childminders and nurseries. After they started school they were frequently left to their own devices, partly because their mother had several relationships over the years, which she kept strictly separate from her children. Tanja's brother was an important attachment figure for her. She often felt she confided more in her brother than she did in her mother. She performed well at school and was a conscientious, reliable student, popular with her classmates. She went to a swimming club three times a week, and really enjoyed training. Tanja had to take on responsibilities around the house very early on, doing the shopping, cooking and cleaning. Added to her homework, the chores became a heavy burden for her. As a teenager, she clashed frequently with her mother. Tanja became overweight and increasingly withdrawn. When she finished school, she followed in her mother's footsteps and trained as an administrator, then moved to another town and looked for a new job. She had been desperate to gain her independence, but she quickly began to feel very lonely. At this point, Tanja had never had a serious relationship. In her new job, she fell in love with her boss and started an affair with him. Her boss was married, and never showed any signs of getting divorced and entering into a proper relationship with Tanja. Tanja felt used, but didn't have the strength to end the relationship and look for another job – especially because she still felt very lonely in this new town. She began to lose weight and finally become anorexic.

The illustration below shows Tanja's profile of basic needs in childhood and at the age of twenty-two. The large areas of basic needs that remain unsatisfied reveal why Tanja has found herself in serious distress and is suffering from anorexia.

Tanja's basic needs and her ability to satisfy them

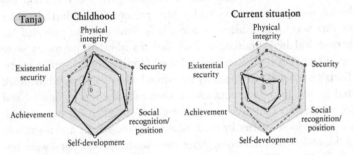

Tanja's basic needs in childhood and in her current situation. Dotted line: Tanja's basic needs; solid line: needs satisfied by the environment.

Physical integrity: Tanja was a healthy child. As a teenager she was overweight, and then became bulimic, eating large quantities of food and usually making herself sick afterwards. She lost more and more weight and finally became anorexic.

Emotional security: Tanja's whole childhood was marred by her mother's unavailability. The constant turnover of attachment figures, grandparents, various childminders and after-school club supervisors added to her insecurity. Her lack of emotional security was felt even more keenly after she moved away. She hoped to find it with her boss, but that didn't happen; her unhappy relationship only served to reinforce her sense of abandonment. Her most important attachment figure was her brother. Tanja has felt even lonelier since Aldo moved to California for an IT internship. They Skype each other every week.

Social recognition and social status: As a child, Tanja received plenty of recognition from attachment figures, such as teachers, and from people her own age. She had two good friends from pre-school right through to sixth form. One of these friends, Saskia, did her training with the same firm as Tanja. But when Tanja moved away they lost touch. Tanja tried to make new friends, but found it difficult.

Self-development: Throughout her childhood and teenage years, her swimming club was everything to Tanja. She particularly loved high board diving. She was talented and took part in competitions. Since moving to a different town she swims only occasionally, on her own.

Achievement: Her performance at school was always adequate. She completed her apprenticeship successfully. But she isn't satisfied by the work she's doing in this new town. It's monotonous and doesn't mean very much to her.

Existential security: As a child, Tanja was acutely aware of her family's financial hardships. After her apprenticeship it was very important to her that she had her own secure income and was able to afford little luxuries from time to time.

When she moved to a different town and started a new job, Tanja found herself in a serious Misfit situation. Even in childhood, Tanja received little emotional security and affection. Her need for emotional security only became greater in her unhappy relationship with her boss. She also lacked social recognition in her private life, and so became increasingly withdrawn. Tanja was frustrated by a sense that she couldn't develop any further and was doing a job that gave her no satisfaction. She began to feel physically unwell and started to lose weight.

Tanja was also unable to use her competencies as much as she wanted.

Tanja's competencies and how she has been able to develop and use them

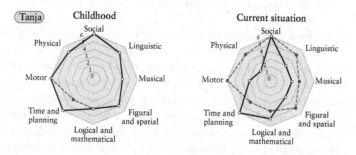

Tanja's competencies in childhood and in her current situation. Dotted line: Tanja's competencies; solid line: the demands of her environment.

Social competencies: Tanja has well-developed social competencies, which made it easy for her to interact with other children at school, and are a great help in working with colleagues in her current job.

Linguistic: She also has good linguistic competencies. But Tanja feels she isn't being challenged enough at work. Writing standard letters and formulaic reports bores her. In her free time she reads literature and books about women's lives.

Musical: These competencies are averagely developed. Tanja listens to country and gospel music. She doesn't like going to gigs on her own, though she enjoyed live music as a teenager.

Figural and spatial: These are extremely good. Tanja would like to do more creative work. She is thinking about taking a pottery class.

Logical and mathematical: Her logical and mathematical competencies are about average. All the same, Tanja finds computerized bookkeeping difficult. She occasionally takes work home with her so she doesn't fall behind.

Temporal and planning: As a child, Tanja felt overwhelmed by household chores. Preparing schedules for colleagues in branch offices takes a lot of effort, particularly as she doesn't feel she gets enough support from her boss.

Motor: Tanja's motor competencies are very well developed. She would like to start using her motor skills properly again.

Physical: Tanja was a pretty little girl. Today, her unattractive appearance upsets her. She doesn't like it when people look at her critically or – even worse – don't take any notice of her at all.

Tanja's profile of competencies, like that of her basic needs, is fairly well balanced. Most values are about average (4), and she achieves her highest values (6) in the areas of social and motor competencies. As a child, Tanja was able to develop her competencies sufficiently. But even then, her limited temporal and planning competencies (3) meant that some tasks overwhelmed her. In her current situation, Tanja feels under-challenged in some areas (linguistic, figural, spatial and motor) and over-challenged in others (logical and mathematical, temporal and planning, physical).

Tanja has given little thought to what she wants her life to look like in the future. Her one aim was always to leave home and become independent. Now she feels a great emptiness inside and wonders what it is she wants to achieve and what is important to her. Tanja starts to work through the things that have happened in her life with a psychologist. She looks at the reasons her mental and physical well-being have been so badly affected, why her sense of self-worth and self-efficacy have reached a low point, and why she became anorexic. After a few months she has a better understanding of herself and the Misfit situation she is in. But she is still finding it difficult to see what changes she needs to make.

Why do we often find it such an effort to make a change in our lives, even when we think it necessary? One important reason is that our basic needs, competencies and ideas are to some extent fixed by the good and bad experiences we have had in the past. We don't just need insight: we also need new positive experiences in order to resolve Misfit situations and recover our inner equilibrium. And this is exactly what now happens to Tanja. One evening she is out shopping and bumps into Saskia, the childhood friend she has lost touch with since changing jobs. They go out for dinner and spend hours chatting. From then on they see each other regularly. Saskia has moved to the same town and has a large circle of friends, to whom she introduces Tanja. Tanja's mental and physical health begins to improve. Eventually she feels strong enough to end the relationship with her boss and find a new job. The work there isn't very challenging either, but her colleagues are nice. In the staff canteen she gets talking to a physiotherapist who tells Tanja about her work. Tanja is excited by what she hears and decides to train as a physiotherapist too. It's a career where she will be able to use her motor abilities and come into close contact with other people. A new network of relationships and positive plans for the future gradually improve Tanja's state of mind. She begins to eat more normally.

As Tanja's story shows, resolving a Misfit situation doesn't have to involve changing our basic needs and competencies – which isn't something we can really do, in any case. We just need new experiences that create greater harmony between ourselves and our environment. People very rarely manage this alone; they generally need support from others. Saskia devoted a lot of time to Tanja and helped her in a number of different ways. Tanja was able to discuss her worries with her, and her new circle of friends meant she no longer felt alone.

Misfit constellations like Tanja's, which are partly rooted in childhood, don't have to dog us for the rest of our lives. Many constellations resolve themselves as we get older, or can be worked through in psychotherapy. But what we find most helpful are positive experiences – which also made a difference in Karim's case. His slight motor disability meant that he had difficulty writing throughout his school career, and suffered badly because of it. During his business apprenticeship he learned to type on a PC and to his surprise was just as quick as his colleagues. For the first time he felt he wasn't inferior, which drastically improved his sense of self-worth and self-efficacy.

Understanding ourselves and others

When things are going very badly and we find ourselves in a serious Misfit situation, we start to take a long, hard look at ourselves and our lives. What is missing from my relationship? Why am I unhappy at work? But people who are by and large doing well may also want to gain a better understanding of their basic needs, competencies and ideas – and the Fit and Misfit situations they have experienced in their lives. Tanja comes to the conclusion that her need for emotional security is much greater than she had previously thought and that she has been neglecting her motor and social competencies. When we understand ourselves better, we feel more in control of our lives. Anyone who thinks that this kind of introspection is unnecessary should bear in mind that we all have ideas about ourselves and our environment. The question is: how realistic are these ideas? And, if they're not realistic, could they actually be contributing to the Misfit situation?

Getting at the truth of our own basic needs, competencies and ideas isn't easy. The following overview offers a brief introduction to how you can approach this task. For example, you might consider whether you felt accepted by your teachers and fellow students at school, or whether you found maths easy or difficult in comparison to other children. This makes us aware of what we were like as children, but also of what we wanted to be like. Other people's views can be helpful here: have conversations with parents, teachers and former classmates, or with friends and work colleagues. Their assessments can be more accurate than our own, because they bring out things that we have long since forgotten or repressed. The support of an expert, such as a psychotherapist, can also be helpful. A useful technique to help you understand a Misfit situation is 'focusing'.[4] You sit in front of a blank sheet of paper and write down key words that occur to you about a particular emotion, such as a nagging sense of unease. The emotion serves as a kind of pilot through the unconscious. You go on listing words until the unease has dissipated a little. Then you have at least captured some of the essential aspects of the emotion.

You may find it helpful to work through the areas in the following overview, creating separate analyses for childhood, adolescence and adulthood, and then ask yourself what has changed over time.

Basic needs

How are my basic needs configured? (Estimate on a scale of 1–7; see Appendix for blank profile to complete.) Which basic needs are particularly important to me, and which less so? Why can I satisfy some basic needs and not others?

- *Physical integrity:* How physically strong did I feel as a child? Was I ever seriously ill? Did I have any accidents? Am I suffering from the long-term effects of these? Do I feel fit and healthy today? Why not? How do I cope with that?
- *Emotional security:* Was I a child who needed a lot of emotional security? Did I feel secure and accepted, in my family, among people my own age and at school? How much affection did I need in comparison to my siblings and other children? Do I get enough affection today from my relationship and my circle of friends? Do I occasionally feel neglected or left out, and why? What am I contributing to this situation, and what is my social environment contributing?
- *Social recognition and status:* As a child, did I get enough recognition from my parents and other attachment figures, or did I experience rejection? Why? What was my social status in

the family and at school? Do I get enough recognition now in my relationship and circle of friends? What social status do I have at work? Am I a loner or a team player? What does my social environment look like in the various different areas of my life, and what is my relationship to it?

- *Self-development:* How did I develop as a child in comparison to my siblings and other children? Was I an early or a late bloomer? Was I able to develop my competencies at home and at school? What hampered my development? Which competencies do I want to develop further? Does my environment support me in this? What are my strengths and weaknesses, and how do I and others deal with them?

- *Achievement:* What things did I enjoy achieving as a child, and what tasks overwhelmed me? What were my grades like at school? What was my level of achievement in comparison to my classmates'? How did I react to academic success and failure? Am I satisfied with my performance today? What competencies do I use to achieve things? Can I keep up with the performance of others? Do I feel under- or over-challenged at work, and why? What contribution is the environment making to this? Why do I have such high expectations of myself? Do I actually have the necessary competencies to achieve the things I want?

- *Existential security:* How did my parents make a living? Did I suffer from material deprivation as a child? What fears and hardships did I take on from my parents? How do I feel about my current situation with regard to income and property? What do my living conditions contribute to this? What expectations do I have of the future?

Competencies

How are my competencies configured? (Estimate on a scale of 1–7; see Appendix for blank profile to complete.) This should be more than just a subjective estimation. Comparisons are helpful – with former classmates when estimating mathematical competencies, for example, or with colleagues when estimating levels of achievement. How do I use my competencies? When do I feel under- or over-challenged? And is that due to my competencies and/or the demands of my environment?

- *Social:* How competent was I as a child, interacting with other children, and how good are my social skills as an adult, within my family or in the workplace? How well do I understand other people's thoughts and actions? How good am I at empathizing, and how willing am I to care for others?
- *Linguistic:* As a child, how good was I at expressing myself in words? Did I enjoy writing essays at school, or did I find it difficult, and why? Do I find it easy to produce a piece of writing today? What are my foreign-language skills like? How good am I at expressing myself in conversations or presentations?
- *Musical:* How good was I at singing and dancing as a child? Did I like music lessons? Do I enjoy listening to music? What about singing and dancing now? Why don't I play an instrument?
- *Figural and spatial:* Did I enjoy drawing and making things as a child? How good is my sense of direction, when driving for example?
- *Logical and mathematical:* Did I find maths easy at school? Do I have an overview of my income and outgoings? Am I good at constructing an argument? How computer-literate am I?
- *Temporal and planning:* Was I disorganized as a child, in completing schoolwork or packing my school bag, for example? How good am I now at organizing my everyday life? Am I capable of making long-term plans?
- *Motor:* As a child, how skilled was I at climbing trees, skipping and playing ball games? What are my gross and fine motor skills like now, in activities such as knitting or tennis, for instance?
- *Physical:* Was I a pretty child who attracted admiration, or was I often overlooked? How do people react to my appearance today? Do I feel comfortable in my own skin?

Ideas

In approaching our own ideas, it is helpful to bear our competencies in mind – figural and spatial, for example – and also to think about the ideas we have in the different areas of our lives, such as our political beliefs.

- What kind of books did I read as a child, what films did I watch, and what interested me about them? What values have I inherited from my parents and other attachment figures, such as teachers? Who were my role models?
- What ideas are important to me today, and what ideas do I try to live by? What expectations do I have of myself, and what expectations are placed on me by other people?
- To what extent are my ideas an expression of my basic needs, competencies and past experiences?

General state of mind

Your state of mind can best be captured by recalling specific situations and asking yourself how you felt and why – during lessons at school, for example, or in the playground.

- How do I rate my wellbeing, my sense of self-worth and self-efficacy over the course of my life to date? When have I felt happy, and when have I felt unhappy?
- Which experiences have improved or damaged my wellbeing, sense of self-worth and self-efficacy? Which situations in life make me happy and ready to move mountains, and why? When do I feel unhappy, helpless and powerless, and why?
- What do I contribute to my wellbeing, and what do other people (my family, for instance) contribute?

Reactions to distress

The best way to understand your own reaction to distress is by thinking about the situations in life when you have been full of energy – situations that are now pleasant memories – and other situations in which you suffered from distress.

- As a child, how did I react when I felt happy or unhappy, when I succeeded in something and when I failed? Did I get sick? Did

I become withdrawn? Did I seek solace from parents or other attachment figures?

- How do I react to Misfit situations today? What are the early-warning signs when I am heading for a Misfit situation? Do I sense when my mental and physical wellbeing is impaired? What mental and physical symptoms do I experience when this happens?
- How do I deal with symptoms such as sleep disorders or stomach aches? What helps, and what doesn't?

Fit and Misfit constellations

You can think through past Misfit situations by asking yourself which of your basic needs were affected and which competencies proved inadequate. What did I do to get myself into this situation, and what was the environment's contribution? You can learn just as much from Fit situations: which basic needs have I been able to satisfy well, which competencies have I used successfully, and how did these things come about?

Self-assessment of basic needs, competencies and ideas, general state of mind, and reactions to distress, Fit and Misfit constellations.

- When have I been happiest, and when were my truly awful times? What have I learned from both Fit and Misfit situations?
- What did my Fit constellations look like in childhood and adulthood? Which basic needs were satisfied and which competencies were especially important? To what extent was this due to other people and to the situation I was in?
- What were my Misfit situations like? How did they come about? Which basic needs were affected? How did I manage to overcome these situations? Which Misfits was I unable to resolve, and why not? Who supported me? What strategies have I developed to cope with Misfit situations?
- How popular was I with my playmates and school friends? Who did I most enjoy being with, and why? What did I like doing in

my free time with friends? What made me like one teacher and not another?

- When did I feel happy at school? What was a nightmare, and why? How frequent were my academic successes and failures? How well did my competencies fit with the demands of schoolwork? My favourite and worst subjects? And what about today: do I feel comfortable at work, and if not, why not? Are my basic needs and competencies a good fit for my job?
- How well do my basic needs and competencies, my expectations and beliefs fit with family life? How harmonious is my relationship? What do my partner and I have in common, and where do we differ? Do I have sufficient social and caring competencies to bring up children, or do I need support?
- Do I get enough recognition in my workplace, and what status do I have among my colleagues? How well do I get on with my managers? Can I achieve the things that are expected of me, or am I under- or over-challenged?
- When do I feel independent, and when do I feel things are not under my control? What do I contribute to this, and what does the environment contribute? In which situations am I unable to act freely and make my own decisions, and why? What would my environment have to look like in order for me to live independently?

Getting to know yourself better can be fascinating and liberating, but it is also – to start with, at least – linked to insecurity and painful realizations. We have to question our old ways and habits, and avoid the scapegoats we have blamed for difficult situations before. We often shy away from deep insights because we fear the consequences and aren't yet ready to change our lives. Self-examination is a complicated business, but worth it in the long term. If we succeed in understanding ourselves, the environment and the situation we currently find ourselves in, we feel more liberated and less prone to thinking that we are helpless to prevent Misfit situations.

We should not, however, expect an ultimate realization to hit us like a bolt of lightning and make all our problems dissolve into clarity and happiness. It takes time and patience to accept ourselves as we are rather than as we would like to be. Insight alone, no matter how great, won't bring a Misfit situation to an end. For that, we need new, largely positive experiences, which will encourage us to take further steps and strengthen our wellbeing and self-confidence. The more successful this is, the better reconciled we will become with ourselves, and the faster we will achieve harmony between ourselves and our environment. Opening ourselves up to new experiences is just as important as searching for insights.

We all have opinions – often preconceived opinions – about our fellow humans. If we want to be fairer towards them, it makes sense to think about them as well as ourselves. Are they doing well, or are they stuck in a Misfit situation, and why? To help us gain a better understanding of their behaviour and their uniqueness, we can think about what their basic needs and competencies are and what kind of environment they inhabit. In which areas of their life are they in harmony with their environment, and where do they clash with it? How do they react to stress? These considerations can help parents understand their children and teachers tune into the idiosyncrasies of students who are playing up in class. In the world of work, it can help managers allocate to junior staff the kind of work that will enable them to use their individual abilities, doing things that are useful and meaningful to them as well as to the company. And when we think seriously about other people, we sometimes experience a positive side-effect: we also learn more about ourselves.

REASSURING PEOPLE AND RETHINKING THE ENVIRONMENT

Coming together

The whole idea of compassion is based on a keen awareness of the interdependence of all . . . living beings.

Thomas Merton (1915–68)[5]

Most people have a sense of social responsibility, and care about their fellow humans. They are pleased when children grow up happy; they are sympathetic when things are going badly for someone and want to help them feel better. They put a huge effort into bringing up their children or looking after older members of the family. No other living creature devotes so much energy to taking care of others of its kind. Over the course of evolution, we have developed pronounced empathetic and caring behaviour for this purpose (see Chapter 5).

The French Trappist monk Thomas Merton – whose philosophy chimes with that of the Fit Principle – tells us that real compassion consists of understanding the way people are tied to their environment. Take the example of a three-year-old child who whines and doesn't want to play on his own. If his parents recognize that their son is envious of his younger sibling, they can increase his well-being by giving him more attention and involving him in activities like changing the baby's nappy. Or an eight-year-old boy who has no interest in playing the piano. If his parents let go of the idea that playing an instrument should be part of everyone's general education, they will stop sending him to music lessons and let him join a football club instead, where he can play to his strengths. Or the son and daughter who look after their elderly father. They can make their father's life easier by understanding that his grasp of space and time has been reduced to the level of a small child's. Their father needs a contained, consistent environment and care that takes account of the fact he can only comprehend short spaces of time.

We always expect other people to do their best. If they don't fulfil our expectations and we find it difficult to accept them as they are, it is often due to our own fears and interests. Parents put pressure on their child because they worry that her poor academic performance will stop her going into the high-status career they have imagined for her. If we get to know people better, we usually find that they may well be doing their best, but various internal and external factors are preventing them from fulfilling expectations. We shouldn't patronize or micromanage them; we should allow them as much independence as possible. Everyone, from the whining toddler to the elderly grandfather, wants to live a self-determined life. We can support them best by reinforcing their strengths, accepting their weaknesses and helping them to inhabit their environment in a way that makes them feel comfortable.

Offering the kind of support that genuinely helps others is a challenging task – take retired people, for example. Some find they have a dramatically reduced sense of self-worth when they stop working. They feel useless because they're no longer achieving things, and miss the social status they occupied in their workplace. We should avoid giving advice – even the kind of advice that we would find useful ourselves. It can easily make people feel that others are controlling their lives. If they ask us what they should do, we should give an opinion, but without insisting that they share it. The most important things we can give to people in difficult situations are compassion and trust. We can be there for them and reassure them that they will resolve the issue in their own way, and that we will give them any support they need. People have to take their own route through good times and bad, and treading that path with them can become a thoroughly satisfying and fascinating experience. It helps us learn to deal with the diversity among people, to accept their differences and let go of false expectations and ideas of normality.

Why we find Misfit situations so difficult

Did our ancestors think about themselves and the world as much as we do? I suspect not. When they got into difficulties, they probably relied on their gut instinct for the most part, and held lengthy discussions in the community about how to help the person affected. We still rely on our gut instinct in our everyday lives, though it often doesn't help us. We also talk to people who share our concerns, though we increasingly turn to professionals such as psychotherapists or advice centres for help.

A life without constant stress and insecurity has become an unreachable ideal. One significant reason for this must be that the (largely unconscious) strategies for coping with life that our forebears used in their small, familiar communities aren't suited to coping with Misfit situations in today's society, with its highly complex structure (see Chapter 7). Another significant reason is that our network of relationships has gradually become smaller and less sustainable. Increasing numbers of people feel they are on their own. But many Misfit situations can only be resolved in a community of familiar people; no matter how competent the psychiatrist or how good the advice centre, they can't help us on a more permanent basis. We therefore have to take a long look at both ourselves and our environment. In Chapter 10 we will explore why it is that we are less and less able to satisfy our basic needs, and how we should reshape our environment so that as many people as possible can lead a fitting life.

10

Changing Times

We need to think the impossible.

> *The modern Sisyphus*
> *Since time immemorial, Sisyphus has been pushing a boulder*
> *up a mountainside, only for it to roll back down into the val-*
> *ley. One day, Sisyphus manages to heave the boulder all the*
> *way to the summit. Since then it has been rolling down the*
> *other side, with Sisyphus chasing after it.*

It is a beautiful autumn day. The air is clear and a light, fresh breeze is blowing. The sunlight is still warm. I am sitting – as I so often do – on a bench in the Münsterhof in Zurich's old town and watching people as they stroll across the square or stand around chatting in groups. A family approaches. While the parents sit down beside me on the bench, their son and daughter skip around the fountain and bound over the cobbles. I enjoy watching them, and tell the parents they have wonderful children. Their eyes light up, and the mother says that they both feel their children are the most important thing in life. And so we fall into conversation.

The father has a senior position in a bank. He is proud of Zurich's financial centre. It's one of the largest in the world, he says. And according to the latest survey, Zurich is one of the best cities in the world for life satisfaction. The mother chimes in: well – for the moment it is. She's a social worker. More and more people are living in poverty and suffering social exclusion in this wealthy city, she says. And most people, even the richest, are under constant stress at work and at home. Even children are being put under unbearable pressure to perform at school. That's true, the father agrees – the future doesn't look quite as rosy as it used to. In the next few years, his bank is planning to make several thousand people redundant. And, the mother adds, even people who are still doing well are starting to feel insecure and worry about the future. Suddenly the warm sun vanishes and there is a chill in the air. We say goodbye and wish each other all the best. On my way home, I start to mull things over.

Are we not like the modern Sisyphus? We have done something no other species has ever managed. Over the whole of evolution, all living things that have populated the earth, without exception, have been at the mercy of nature. None of them has been stronger than its environment. We are the first and so far the only creatures to control the environment with our discoveries and inventions, and now – in an ironic twist of fate – we are starting to feel controlled and even threatened by our self-made environment. We are chasing runaway progress and feeling fearful of what our lives will look like in the future.

But to halt this train of thought at once: I am optimistic – quite simply because in the past 200,000 years humans have always found ways of overcoming even the most difficult situations. In this final chapter, I would like to put forward a handful of thoughts about Fit and Misfit on the macro scale. I want to look at how society, the economy, family and communities need to be reshaped so that we can live a fitting life. In order to do this, we have to be prepared to think the impossible: a universal basic income, for example, and new forms of family and community living.

GENERAL INSECURITY

People in the Western world have never had it so good. No one is starving; in fact, excess weight is becoming a worry for some. People are looked after by well-developed health and welfare services. Their children attend good schools. They live under the rule of law and feel their lives and property are well protected. Very few people want to turn back the wheel of time. So why are we so insecure?

'Too big to fail.' In 2008, we went through a global financial crisis, and banks were rescued in a panicked operation that cost the USA more than the whole of the Second World War. People began to wonder how this disaster could have come about: economists hadn't been able to predict it – or perhaps they just didn't want to. Why were none of the banks that were to blame for the crisis held to account? And if everyone had in fact been acting in good faith, then what was to stop an even larger economic catastrophe happening in the future? Globalization has turned the economy into a monster steered by forces that, as we saw with the financial crash, are ever more difficult to keep in check. The economy has become too big and too complex to be controlled.

People are also losing faith in the state. Rightly or wrongly, some sections of the population are under the impression that our politicians care less about the good of the nation than about their own interests; they are influenced by all manner of lobbyists, and caught in a web of political and industrial interest groups. Many people blame the 'elite' for all kinds of crises. They increasingly doubt that politicians are still capable of dealing with the problems we are facing, such as unemployment and an ageing population. Their impression of politicians comes from the media, and they can't see what these people are actually achieving. Fewer and fewer voters are turning out, because 'the elite are going to do whatever they want anyway'. Our unease has become so great that, as with the economy, there is a growing conviction that, however hard they try, no politician can get a real overview of such a complex society and lead the country competently. 'Too big and too complex to govern.'

The fact that they no longer understand society and the economy creates huge anxieties in people's minds. They wonder who is actually responsible for things like the security of the internet's immense databases and global communication networks. And when they can't find answers to these questions, they adopt the same attitude they have taken to nuclear power plants: they capitulate before the complexity, but they are aware of the danger. They can only hope there is never a worst-case disaster that would plunge millions of people into misery. It is easy to imagine a cyber-attack causing a total power cut lasting days or weeks that would plunge us into a social catastrophe. A power cut like the one Marc Elsberg describes in his novel *Blackout* would cripple the food supply chain, the health service, communication, transport – everything, in short.[1]

People are feeling insecure because they depend on the state and the economy in an existential sense (see Chapter 7). State institutions support us throughout our lives, from nurseries and schools to the health and welfare services, to retirement and care homes. The economy enables people to earn money and have a good quality of life. Both citizens and politicians are increasingly worried about whether we'll be able to finance state institutions in the future, and economists doubt that the private sector can go on providing enough jobs. It comes as no surprise, then, that more and more people are suffering from existential angst and worrying about their children's future.

But people's insecurity isn't just existential: it is emotional and social as well. They are feeling less and less emotionally secure and increasingly lonely. They lack the emotional security and social support that family and community used to provide. Although we do still live in family-like structures, these have shrunk and become less stable, and today's communities also provide less of a refuge of emotional and social stability (see Chapter 8).

If we lose our faith in society and the economy, we are also liable to drop ancient patterns of behaviour and ideas that stem from the very earliest communities. And this is something that isn't just happening in a community of a few hundred people, but in a society of millions. People are starting to isolate themselves and exclude others. They are becoming susceptible to movements on the far left and (a particular problem at the moment) the far right of the political spectrum. Populists are stoking their fears, making unrealistic promises and holding out to them the sense of 'us' that they are so desperately missing. This kind of thinking can unleash enormous social and political forces that populists and ideologues can abuse, and it can ultimately lead to violent conflicts. Politicians and journalists blame a huge variety of factors – often ethnic, cultural and religious – for the unrest. But the things that really lie at the root of these ills are people's collectively unsatisfied basic needs.

Society and the economy have become runaway successes that are now at risk of going off the rails. And that would be disastrous: for better or worse, we are reliant on an intact state and a functioning economy to satisfy our basic needs. It isn't just the size and complexity of society and the economy that make people feel insecure. Above all, it is the reasonable fear that they will no longer be able to fulfil their basic needs, and the sense of powerlessness that comes with that fear.

FIT AND MISFIT IN SOCIETY AND THE ECONOMY

In the nineteenth century, people started building zoos all over Europe. Exotic animals from every corner of the earth were locked up in cages to be marvelled at by visitors. Over the course of the twentieth century, zookeepers began to realize that these animals were unhappy. The panther paced back and forth in his cage all day; the bears were tearing out their own fur and the polar bear refused to eat. The keepers started to think: what do these animals' natural habitats look like? How do they feed? Do they live alone or in groups? What experiences do they want to have – do they want to search for food, for example? The keepers created living spaces for the animals to fit their needs. They gave the antelopes a steppe-like landscape with enough space to run; there were trees for the bears to climb; hollow logs filled with fruit and berries to challenge the chimpanzees before they could get at the food. The animals could choose to do these things whenever they wanted. And, in recent years, people have finally come to the conclusion that certain animals should no longer be kept in zoos, because we cannot do justice to their needs.

We are far more adaptable than any other animal, but we still depend on a 'species-appropriate' environment. We no longer live in our original habitat, in small, manageable communities with people we know, surrounded by the natural world, but in a society that has only existed for around 200 years. The blink of an eye in comparison to the 200,000 years that our ancestors survived in their small communities. The number of people who make up a community has also changed dramatically. Our ancestors' communities comprised a few hundred people, and modern societies are composed of many millions. This fact has fundamentally changed our interpersonal relationships. In a small community, everyone knew each other and had a fixed social position. In our society it's impossible to know everyone, and fixed social positions have largely been lost. Small communities were constituted in the same way for centuries, if not millennia; ways of life and shared values were passed down over many generations. Today's society, by contrast, is always changing, and doing so at an increasingly rapid rate.

Our basic needs and fundamental values are still rooted in communal life, so it comes as no surprise that we are finding it more difficult to get our bearings in our anonymous mass society. We are losing trust in a highly complex state and a globalized economy, and grieving for a culture that gave us an identity. Is this environment that we have created for ourselves still 'species-appropriate'? And if not, how can we make society and the economy more human?

The answers to these questions are drawn from the Fit Principle, which lies at the heart of this book: every human being is unique and wants to live a fitting life (see Chapter 8). We will only feel well within ourselves, with a good sense of self-worth and good self-efficacy, when society and the economy are shaped in a way that allows us to satisfy our basic needs sufficiently.

Trust and participation in state and society

People feel insecure because the state and its institutions have become completely opaque to them, and ordinary people have all but lost the ability to help determine society's direction of travel. People are not herd animals, blindly following a lead. Now more than ever, they want to be regarded as mature beings and take an active role in state and society.

To modify the famous quotation from Winston Churchill, recent decades have proven that democracy is the best form of government. It is by no means perfect, but it can be improved at any time. Here is some food for thought on how trust and participation can be increased among ordinary members of society:

A fully functioning state. The best way for the state to build trust is by being functional and transparent. For this to happen, Members of Parliament need to do more than just represent a party line or a political movement; they have to do credible work to allow citizens to fulfil their basic needs, and they have to possess the right competencies – in family policy, for example. The government should not consist of politicians who have worked their way up through the party ranks, but of prominent figures from society with as much expert knowledge as possible in their policy area. As the fourth estate, the media's task is to ensure transparency.

Direct democracy enables citizens to influence politics directly, through referendums as well as elections. Direct democracy, which has been practised in Switzerland for more than a century, is an essential means of involving the population in the political decision-making process, encouraging active public debate and legitimizing political decisions. There is a huge difference between a bill on the naturalization of foreigners being passed by Parliament and the sitting government, or through a referendum preceded by a lively public discussion of its pros and cons.

Delegating power. As nations increasingly become part of trans-national or even global structures such as the EU and the WTO, people start to feel 'homeless'. This effect can be counteracted by passing responsibility and decision-making powers down from the national to the regional level, and from the regions down to communities. In Switzerland, citizens can vote on whether personal and business taxes should be raised or lowered, whether schools or hospitals should be built, and whether cultural institutions should be given more or less financial support. When political responsibility is delegated, people stop feeling as though things are out of their control and start engaging in politics.

In some countries, politicians and governments believe they can manage without their citizens' trust and participation. That may be the case in good times, but certainly not in bad. And by then it's too late.

Living culture

Culture is an essential element of any community (see Chapters 1 and 6). People identify with their community through its culture. The tremendous sense of belonging that comes from culture is demonstrated by the history of Judaism. The Jewish people have undergone the most terrible suffering: captivity in Babylonia and Egypt, countless pogroms in the Middle Ages, centuries of diaspora, the Holocaust in the twentieth century – and yet their identity has survived for 2,500 years. A highly differentiated culture, bound very tightly to religion, has kept the Jewish community together. Something else we can learn from Jewish culture is that in order to survive a culture must remain current, with people participating in many different ways but guided by common values. This is something that modern society has lost sight of. In the space of a few generations, cultures that have survived for centuries have given way, as high cultures have succumbed to a globalized entertainment culture and universal values. A whole range of factors have contributed to this state of affairs, among them worldwide communication tools such as social networks, increasing mobility, the disappearance of regional customs, traditions, dialects and even languages, and the loss of a historical consciousness. This cultural transformation has led to a gradual loss of identity. Taken in isolation, feeling like a citizen of the world doesn't create an identity. Real culture has limits; it is restricted in geographical terms so that it can be kept alive through a constant exchange between people who know each other. This is why we need new ways of living together, as I will outline in the section on family and small communities below.

Education is a vital cultural asset. The education system must not simply consist of a collection of institutions producing competent workers for society and the economy. The cultural task of these institutions lies in encouraging individual development all the way from pre-school to university, and helping to shape people's idea of humanity and sense of community.

A future-proof economy and meaningful work

It was the start of a social development unprecedented in human history: having been self-sufficient for 200,000 years, in the space of just 200 years the majority of humans stopped providing for themselves directly. Today, most people receive an income from an employer and use it to purchase accommodation, energy, mobility and access to a wide variety of services and a global marketplace of consumer goods. They no longer have to worry about their own subsistence – something

Figure 10.1: Working the fields in Africa (*left*); industrial agriculture in Brazil (*right*).

that was a great burden for their ancestors, but also gave their lives meaning. The price we pay for this is being heavily dependent on a salaried job, and doing work from which fewer and fewer people can derive any meaning. And yet we still have a huge need for existential security. What worries us most isn't air pollution, the destruction of the rainforests and global warming, but a potential economic collapse that would damage our wellbeing.

Following recent economic developments, it's easy to understand why people are suffering increasingly from existential anxieties, especially fears for their children's future. In Spain and Italy, 20–30 per cent of young people are unemployed, and in Greece the figure stands at over 40 per cent.

The transfer of manufacturing to low-pay countries such as Bangladesh and China has led to a loss of jobs. But the greatest reduction in jobs has come from the progressive automation and digitization of work processes. It started with manufacturing work – in the automotive industry, for instance – being taken over by machines and robots. The economists Michael Hicks and Srikant Devaraj have shown that 86 per cent of job losses in US industry between 1997 and 2007 were down to the digitization of production methods.[2] In the future – and to some extent, this has already started to happen – we won't just see workers who perform monotonous tasks at conveyer belts and supermarket checkouts being made redundant, but qualified professionals including bus and train drivers. They will lose their jobs with the introduction of driverless vehicles. Digitization is also starting to encroach on the service economy, where three quarters of all employed people work. The routine tasks performed by sales staff are gradually being taken over by computers, and the advice given by insurance agents and bank staff provided by internet platforms.

Economists predict that job losses will be high over the coming decades. A detailed study from the UK in 2013 came to the conclusion that 47 per cent of all jobs in the USA were under threat from digitalization.[3] A study by McKinsey claims that we already have the technology to automate 45 per cent of all work currently performed by humans.[4] And there is no stopping this development, for the simple economic reason that machines and robots are cheaper than human workers.

If we really did end up facing mass unemployment, the welfare state would be completely overwhelmed. Social tensions would immediately increase, our already fragile sense of solidarity would collapse, and we might see social unrest of undreamed-of proportions. And so we must take a serious look at how our existential security and the maintenance of the welfare state can be guaranteed in future.

One possibility currently under discussion is the universal basic income. All citizens, regardless of their situation, would receive financial support from the state. They wouldn't have to do anything in return, and could top it up with paid work. The economist Thomas Straubhaar suggests that the basic income could be financed by taxing every other form of income – not just salaries, but interest, distributed profits, dividends, royalties, rental income and the profits from transactions and speculation.[5]

It isn't just income, but assets and property that contribute to our existential security and quality of life, social respect and status, and the exercise of power. In the past, acquiring property was a successful evolutionary strategy for improving a person's chances of survival. But as property was never guaranteed to remain in our possession, we began to pursue it obsessively. Even someone who is a billionaire several times over, the proud owner of a financial empire with villas all over the world, a private jet and a yacht, still wants to get richer. What was once a meaningful survival strategy has become a disaster: the social and economic structures in which we live have allowed a few people to accumulate a disproportionate amount of property, while the majority sinks further and further into poverty.

The total sum of all financial assets worldwide, including real estate, is estimated at more than $200 billion.[6] In the last decade, this unequal distribution of wealth has led to increasing dissatisfaction in the general population – and not just in developing countries, but in prosperous ones, too. The income and wealth of the super-rich has increased sharply, while for the vast majority wages have stagnated or even fallen:[7] 0.001 per cent of the world's population (around 90,000 people) controls more than 30 per cent of global assets; 0.01 per cent (around 800,000 people) owns a total of 62 per cent; and 0.1 per cent (8 million people) owns 81 per cent – 99.9 per cent of the nearly 8 billion people on the planet share the remaining 19 per cent of global wealth.[8]

For a long time, neoliberal economists have defended this skewed wealth distribution with the argument that, in the long term, it benefits all layers of society, as wealthy people invest in the economy and generate jobs and higher incomes. And for more than 100 years that was true. But the 'trickle-down effect' has stopped working. Wealthy people prefer to speculate on the stock exchange and invest in profit-making companies such as Minera Escondida in Chile. The largest copper mine in the world is 58 per cent owned by an Australian–British company, 30 per cent by a British company, 10 per cent by a Japanese company and 2 per cent by an international finance corporation. The profit doesn't go to the people of Chile; it goes to mining companies and traders in raw materials based in Geneva, New York and Singapore, who hide their profits in offshore companies in order to pay as little tax as possible. The local population is left empty-handed, forced to work for miserable wages and live in inhuman conditions. They rightly wonder why firms, corporations and mineral wealth belong to rich people elsewhere in the world and not to the Chilean people. And why the profits they have created go to the shareholders rather than the miners and other employees. In a fair world, every individual would have to work for their income and assets, and wouldn't be able to use disproportionate ownership and economically damaging speculation to enrich themselves at the cost of others.

This exorbitant wealth must be redistributed, perhaps using the fairest tax of all: inheritance tax. Ownership of land must also be rethought. Land, like air and water, should be common property.

Ownership and wealth should still be permitted, but only insofar as other people are not disadvantaged and made dependent by it.

People don't want to work merely in order to earn a living. Work should also satisfy their basic needs for self-development and achievement. People want to develop and to achieve things that satisfy them and have meaning, things that strengthen their sense of self-worth and self-efficacy (see Chapter 4). Work is an aspect of human dignity and contributes to social status in the community. In the past, people often worked independently as farmers, craftsmen and tradesmen. A joiner would design and then produce his own furniture. He was a joiner all his life and constantly improved his craftsmanship. He was proud of his job and his workshop, and he had a fixed social status within his community. Today, most working people spend their lives as employees. Their work largely consists of performing tasks they have been assigned. Self-determination and self-development are a rarity. Modern working conditions mean that people identify less and less with their work and their employers. They also receive very little social recognition for their achievements. People are now less able to satisfy their need for achievement because they are less able to develop and use their unique profile of competencies (see Chapter 5). People who are good at working with their hands and would like to do something physical are being forced to take office jobs to which they aren't suited, leaving their need for achievement unsatisfied. This also robs work of its meaning. Work that people have been instructed to do by someone else, which doesn't suit their competencies, is linked to stress, and this is leading to more people becoming physically and mentally exhausted, and depressed, and in the worst cases succumbing to burnout.

In the globalized, digitized economy of the future, we will be even less able to satisfy our basic needs for self-development and achievement. And that means we need to create new living spaces and communities where people can once again satisfy these individual needs, as well as their needs for social recognition and status.

FAMILY AND COMMUNITY
FOR A FITTING LIFE

Living in family groups and tribes ensured that *Homo sapiens* survived for 200,000 years, and the ancestors of *Homo sapiens* lived in simply structured communities for millions of years before that. It's likely that more than 6 million years ago our most distant ancestors – those we share with primates – lived in groups, in the same way that gorillas and chimpanzees do today. Family and community are an evolutionary legacy that is deeply rooted in our psyche.

To our ancestors, family and community were important for survival. The same is not true for modern humans – but these things are still essential for our mental wellbeing. We can only satisfy our basic need for emotional security and affection within the family, and our need for social recognition, status and a sense of belonging in a community. Our social competencies such as attachment behaviour, caring and interpersonal communication were shaped in families and communities, and aren't suited to life in an anonymous society.

The fragmentation of communal life

In all countries where social, technological and economic progress has taken hold, families and communities have slowly dissolved over a few generations into an anonymous mass society (see Chapter 1). Large families with lots of children and relatives have become nuclear families with one or two children and much more distant ties to other relatives. Partners and parents increasingly live apart, and children grow up in a wide variety of relationship constellations, including small single-parent families and stepfamilies (see Chapter 8). The family is in danger of losing its original emotional and social stability, which is hugely important for children (see Chapter 7).

As with families, a number of factors, including the diversification of society and the economy, have contributed to the dissolution of small communities. The separation of home, family and workplace, along with a rapid increase in mobility, has led to the fragmentation of communal life (see Chapter 7). The effect of this societal development on people's lives can be seen most clearly in how they satisfy their basic needs. Up until 200 years ago, people satisfied their basic needs exclusively within their family and community, which bound them very tightly together (see Chapter 7). Providing food, caring for children and protecting people and property were communal activities. A common language, customs and religious ideas also helped to bind people into the community. This emotional, social and existential cohesion has largely been lost from modern society. Food is provided by an industrial supply chain, and the healthcare system looks after our physical wellbeing. It is still the job of parents to bring up their children, but education is mostly provided by schools. The state (the legal system, for example) and the economy, in the form of employers, are responsible for our existential security. But our emotional and social basic needs – and this is one of the Fit Principle's central insights – can only be fulfilled in families and communities. Neither society nor the economy is in a position to do this.

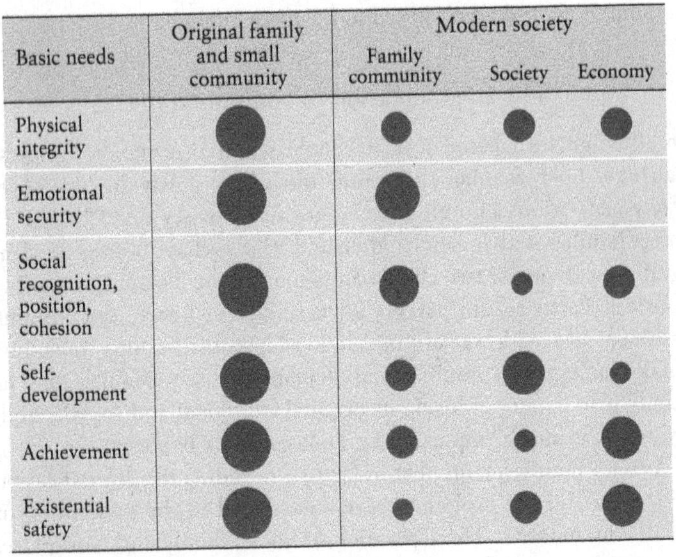

Basic needs	Original family and small community	Modern society		
		Family community	Society	Economy
Physical integrity	●	●	●	●
Emotional security	●	●		
Social recognition, position, cohesion	●	●	·	●
Self-development	●	●	●	·
Achievement	●	●	·	●
Existential safety	●	·	●	●

Figure 10.2: The satisfaction of basic needs is transferred from family and community to society and the economy.

More and more people are finding their wellbeing is impaired by the breakdown of family and community. Consumerism, the entertainment industry and the internet's social spaces are poor substitutes. People are increasingly going to see doctors or psychologists not because they're suffering from an illness, but because their wellbeing has been impaired by the circumstances in which they live.

Why we can't do without family and community

We need some form of communal life suited to the times we live in; communities in which people of all ages can find emotional security, recognition and solidarity, and live as they choose, staying true to their individuality.

In order to develop well, children need a social environment that only family and community can provide. The strong bond between parent and child forms the basis of family cohesion. It is the only relationship in which children feel unconditionally accepted. If they are going to develop a healthy basic trust and grow into emotionally stable and socially competent adults, children need parents who are available and reliable, who react appropriately to their behaviour and above all satisfy their basic needs sufficiently (see Chapter 4).

The family has never been a social island, on which parents bring up their children alone. It has always been part of a community. Alongside their parents, children need relationships with trusted adults who can serve as role models and provide them with the experiences they need for their development (see Chapter 3). They also need relationships with children of different ages; they want to make friends and experience things together. Stable, long-lasting relationships with adults and children are essential to their development.

Some adults manage perfectly well all on their own – or they would, if it weren't for their need for emotional security. In the long term, most people feel lonely and stressed without a minimum of emotional security. Their need may be less acute that it was when they were children, but they still long to be accepted unconditionally. If they fall in love, they experience an all-consuming emotion similar to the parent–child relationship. Unfortunately, that feeling of being in love doesn't last a lifetime like it did for Baucis and Philemon. When people enter into a romantic relationship, they often have such high expectations that they place too much strain on the relationship. A partner is also now expected to serve as a substitute for the community we lack. As with the family, being part of a community is a fundamental ingredient for a viable, lasting partnership. A partnership is disburdened, enriched and reinforced when it is integrated into a broad network of relationships that offers opportunities for experience, but also comes with communal responsibilities.

Older people are another large group whose wellbeing depends on life in a community. And not just because some are less independent and in need of care; they are also increasingly reliant on closeness and affection. They often lack emotional and social security because their relationships with children, relatives and friends have become looser and less binding. As they enter retirement, they also lose their social status; they feel useless and a burden on their family and the state, and this has an additional impact on their sense of self-worth. A small community would give them a circle of familiar people to chat and have fun with. Contributing to the community would also make them feel useful – telling stories to younger children and helping older ones with their homework, for example. A community is essential if elderly people are going to feel they are in good hands and able to lead a meaningful life.

We need to find new ways of living so that people of all ages can feel emotionally secure and have a sense of belonging; so they can receive social recognition and have a secure social status. Only in enduring, trusting relationships are people prepared to accept others with all their strengths and weaknesses, and to support and look after one another in difficult times.

Rethinking family and community

We can't turn back the wheel of time, and nor do we want to re-create the small communities of the past. They were often characterized by authoritarian family structures, and village communities frequently exerted a high level of social control that limited people's self-development. But we have the chance to work out new ways of living together, and making them a reality poses a difficult but potentially very satisfying challenge.

The conditions in which our ancestors lived forced them to form small communities. Communities are no longer vital for our survival, and on the surface of things it looks like we can do without enduring, sustainable interpersonal relationships. We have grown accustomed to a life of great individual freedoms and little interpersonal responsibility, and we are reluctant to give it up. We are in constant competition with each other, always having to prove ourselves as partners and employees, and constantly in danger of dropping out of all our relationship networks and becoming socially isolated. We live an extremely individualistic life, at the expense of solidarity.

The commune is one way of living that promotes both individuality and solidarity. It is formed by a group of people voluntarily coming together after getting to know each other well and making a joint decision to embark on this form of communal living. They might start a housing cooperative, to which they all make a financial contribution. The cooperative could also receive state support – though more of that later. A sense of belonging is fostered through lively discussions and a wealth of shared experiences. People are no longer just passive consumers, but active participants in communal life. They are able to do meaningful work that is suited to their individual abilities.

Mutual give and take and shared interests function as brackets that hold the community together. People support each other in their everyday lives, caring for children and the elderly, pursuing hobbies and playing sports together. Relationships grow deeper when people help each other out, solve conflicts together and take responsibility for their community.

We can see just how much we need a more communal approach to life from the recent increase in communal activities such as allotments, swap shops, repair cafés and inter-generational living projects, founded and maintained by enterprising individuals. Communal living provides the ideal conditions for this need to be satisfied fully and in the long term.

The state supports communal living and profits from it

Communes can only fulfil their task with the necessary infrastructure in place. The community buys property to serve as communal accommodation. The individual blocks of flats and gardens are renovated and turned into communal living spaces. Ideally, a group will buy a plot of land and build a settlement based on the ideas of the people who are going to live there. As well as ecological considerations such as the provision of geothermal and solar energy, the main aim here is to create living spaces in which people feel happy and comfortable. Everyone is therefore consulted during the planning and building or renovation of the property – even the children. How should the flats and communal spaces be laid out? Does it make sense to include removable walls in the design? How can stairs and lifts be designed to suit older people? What should the gardens and play areas look like? Communal planning takes a lot of time and effort, but it results in the desired quality of life and reinforces a group's sense of belonging and shared responsibility.

From the Fit Principle point of view, society's most important task is to provide living spaces in which people can live a fitting life. As part of this solution, the state needs to support these new forms of family and community with tax exemptions for families and good mortgage rates for communal living projects. It could create legal and planning conditions that make it easier to build communal accommodation. Property rights could be adapted and communes could be protected by legal constraints that largely take them out of the property market.

Finally, industry also has a duty. If people are going to live a fitting life in their communities, working conditions need to be improved as well; most importantly, parents need more time off. That means sufficient maternity and paternity leave and flexible hours, as well as part-time working arrangements that don't discriminate against mothers and fathers in terms of income and career progression. This is not a utopia. The Scandinavian countries are showing us the way – and giving people these rights has not brought them any economic disadvantages.

Communal living doesn't just represent an expense to the state: it also relieves the state of some costly tasks that it can't perform satisfactorily in any case. Children and older people, those with disabilities and chronic illnesses, can be better cared for in small communities than they can in state institutions. The care and support a community provides are better value for the state than running playgroups, day-care centres and nursing homes.

Finally, small communities can help the state to solve a serious social problem. Many countries are now caught in a demographic trap. People are living longer, and not enough children are being born to keep the welfare state on an even keel. The cost of care and pensions is steadily increasing, and the number of working people who must pay for these things is falling. An ageing population can only be avoided if the number of children born every year is at least as great as the number of men and women in the previous generation. In statistical terms, this 'replacement' birth rate equates to 2.1 children per woman. Germany and Switzerland have a birth rate of just 1.4. The figure is even lower in all the southern European countries, including Italy, with its history of large families. Each year we miss the target number of births for long-term population stability by over a third. In desperation, countries with very low birth rates, such as Japan and Italy, have called on their people to start families and have more children – but without success. The Scandinavian countries, where women have an average of 2.0 children, show that a 'replacement' birth rate can be achieved. They support families by giving them sufficient parental leave, free childcare and all-day schooling, and they have also introduced social and economic measures to ensure work and family life are compatible.

Communal living for everyone? What about all those who don't want to give up their own home and garden and their beloved 4x4? Those people for whom communal living – in the form outlined here – means too many controls and obligations? They are free to go on living as they have always done. But everyone who wants to live communally should have that option. And they won't regret it. They will bear less of a financial burden, and parents will find it easier to combine work and family life. Their lives will regain some of the meaning that has been lost from our society and economy. They will participate in other people's lives rather than focusing solely on their own wellbeing. They will be pleased for members of their community when things are going well for them, and support and comfort them when they are going through difficult times. They will see children being born and growing up, and people falling ill and growing frail. They will be less reliant on the recognition they receive and the things they achieve in the workplace, and less dependent on material goods, because their community will allow them to live a fitting life.

It was my intention in this chapter to set out the basic elements that might constitute a modern community. A community should allow people to fulfil all the needs they are unable to satisfy in society and the economy. Once the basic principles are established, people can shape their community in their own way.

LIVING A FITTING LIFE IS A HUMAN RIGHT

The Fit Principle applies both to individuals on the micro level and to society on the macro level – to the whole of humanity, in fact. Each of the 8 billion people on the planet is striving to exercise their individuality in harmony with their environment. They want to satisfy their basic needs, develop and use their competencies and live according to their ideas and beliefs. When people succeed in living a fitting life, they feel well both mentally and physically, and have a good sense of self-worth and self-efficacy.

But before they can live a fitting life, people need a society and an economy that will allow this to happen. And not just in the more prosperous countries, but right across the globe. True humanitarianism doesn't just mean sympathizing with the suffering of others and sending emergency aid, but enabling people to satisfy their basic needs independently. All over the world, billions of people are living in poverty and misery, and falling victim to natural disasters and starvation. Every day they see images in the media of the great wealth that exists in Europe and the USA. It is no wonder that more and more people are making the journey there to escape their misery. We must use knowledge transfer and financial support to help them satisfy their basic needs themselves. We need the will to set these things in motion, and the humility and patience to see them through. In the long term, everyone will benefit, including us. Peace is not a utopia; it can become a reality if people are able to lead a fitting life.

Appendix

THE ZURICH LONGITUDINAL STUDIES

The goal of the Zurich Longitudinal Studies was to measure the diversity among children in as much detail as possible and gain a better understanding of the norms of child development. What we learned not only deepened our understanding of children as such, it also helped us to appreciate the individual characteristics of children with developmental and behavioural disorders, enabling us to support their development more effectively and advise parents and professionals in a more child-centred way. This ambitious research project involved a team of paediatricians, development specialists and biostatisticians, who recorded the development of more than 900 children from birth to early adulthood. Between 1954 and 2005, they carried out four longitudinal studies for the Growth and Development Centre at Zurich University Children's Hospital.

The *First Zurich Longitudinal Study,* which involved 350 children, ran from 1954 to 1974 as one of the European Collaborative Studies.[1] The children were assessed at the ages of 1, 3, 6, 9, 12, 18 and 24 months, then annually until the age of ten, and every six months from that point until the age of eighteen. The main aim of this study was to describe physical development as fully as possible. This gave us the largest database of longitudinal growth data in the world, providing data on twenty-two anthropometric measures, together with bone age. The detailed data enabled us to answer a host of scientific questions such as the relationship between growth parameters, such as height and weight, and the dynamics of physical development, such as the appearance of secondary sexual characteristics and growth spurts during puberty.

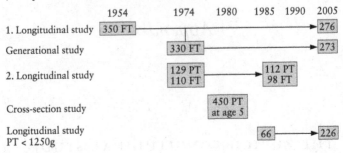

Zurich Longitudinal Studies 1954–2005
(PT = pre-term births; FT = full-term births)

The *Generational Study*, which began in the 1970s, looked at the offspring of people who had participated in the first study. The development of 330 children was tracked from birth to the age of eighteen. This study investigated areas of development, including cognition, speech and motor skills, and behaviour, such as sleeping and bladder/bowel control. We conducted lengthy interviews with the parents to obtain detailed information on the conditions in which the children were growing up. The study allowed us to compare growth and development data taken from one parent and their children.

The *Second Zurich Longitudinal Study* was carried out for ethical reasons. In the 1970s, thanks to huge medical and technological progress in perinatology, more and more preterm babies began to survive. Since these infants had often survived serious complications, we felt duty-bound to carry out careful follow-up checks on them. The study comprised 129 preterm babies and a control group of 110 full-term babies, whose development was tracked in the same way as the children in the Generational Study. Additional data was obtained from a cross-sectional study in which 450 children who had been born prematurely were assessed at the age of five.

The *Longitudinal Study of Severely Premature Infants* was conducted between 1985 and 2005 using 226 children whose weight at birth had been less than 1250 grams, and who had a significantly higher risk of developmental disorders. Their development was also studied from birth to early adulthood.

The results of the Zurich Longitudinal Studies were published in over 120 original journal articles. More than 100 overview articles were passed on to professionals such as teachers and therapists. Eight books on child development came out of the studies, and have attracted a broad readership, particularly among parents. Detailed references and further information on the Zurich Longitudinal Studies can be found at www.largo-fitprinzip.com.

Practical instructions for using the Fit Principle

Below you will find some practical guidance for using the Fit Principle. This guidance is aimed at readers who are interested in using it to evaluate themselves or someone close to them, such as their own child. You should also refer to Chapters 8 and 9, which set out how to assess your basic needs, competencies and ideas, record your general state of mind and your reactions to distress, and recognize Fit and Misfit constellations. The questions contained in the lists on pp. 281–6 are a good starting point for thinking about the various different areas of the Fit Principle. It is worth separating these areas into childhood, early adulthood and your current situation as you work through them, and then asking: what has changed over time? How did these Fit and Misfit situations come about? Below is a set of brief introductions and radar graphs for you to photocopy and complete. It helps to use different colours here.

How are my basic needs configured?

Chapter 4 gives a detailed description of how to evaluate your basic needs. A self-evaluation should not be based purely on subjective impressions: you should always make comparisons with other people. For example, how great is my need for social recognition and social status in comparison to that of my siblings and friends? Other people's views will also help to objectify your own assessment. How high do my parents think my need for emotional security and affection is? How do my former teachers rate my need for self-development and achievement? Each basic need should be rated on a scale of 1–7, and the values entered into the radar graph.

You should also rate the challenges and expectations of your environment for each basic need on a scale of 1–7 and record them on the diagram. For example, does my family think I pay too little attention to my existential security? Does my employer expect me to perform better than I do?

Fit and Misfit constellations can be seen from the degree to which your self-evaluation corresponds to the levels entered for the environment. A wide gap between the two numbers entered in the section for social recognition and status, for instance, reveals a Misfit situation. You can then ask the question: how did this come about? Am I responsible for this situation, is the environment at fault, or are both of these things true?

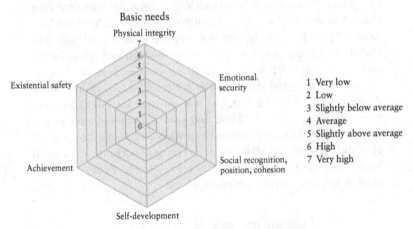

What are my competencies?

Chapters 5, 8 and 9 give a detailed description of how to evaluate your competencies. Your self-evaluation should always take into account comparisons with other people: in the area of mathematical competencies, this might be former classmates, or in temporal and planning competencies, it might be colleagues. Other people's assessments will help to make your own more objective – ask parents and teachers when evaluating your linguistic competencies, for example. Each competency is rated on a scale of 1–7, and the values should be entered into the radar graph.

You should also look at the degree of harmony between your individual competencies and the challenges and expectations of the environment. Estimate these environmental factors on a scale of 1–7 and enter them into the radar graph. What demands, for example, does my employer make on my linguistic competencies?

Fit and Misfit constellations: the degree of correspondence between your self-estimation and the environmental demands will point towards Fit and Misfit constellations. As an example, a difference between your self-estimation and the expectations and demands of your workplace might highlight a Misfit situation in the area of social competencies. You can then ask the question: how much potential do I have in this area, and where are my limits? Why do I feel my social competencies are being overstretched or under-utilized?

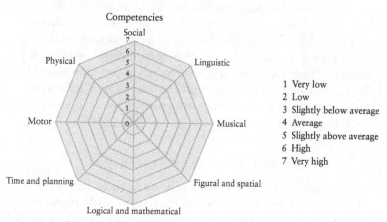

1 Very low
2 Low
3 Slightly below average
4 Average
5 Slightly above average
6 High
7 Very high

What do my ideas look like?

When you come to evaluate your ideas, you may want to refer to the detailed exploration of ideas in Chapter 6. Your evaluation should include ideas that occupy all humans, such as self-image, family and marriage, upbringing and school, morality and religion. Then there are ideas that are important to you as an individual, such as environmental awareness. Make separate notes on each idea. These should also be rated on a scale of 1–7, with separate entries once again for childhood, early adulthood and current situation. Asking questions of your parents and friends, for example, helps to make your notes more objective.

You should also consider the extent to which your ideas correspond to those that are prevalent in your social environment. You might ask: what political opinions do my parents hold? What are my colleagues' views on the economy – the causes of unemployment, for instance?

The degree of correspondence between your own ideas and those of the environment point towards Fit and Misfit constellations that crop up over the course of your life, but which can also disappear again. Children mostly share their parents' ideas on morality and child rearing, but as adults they often form different opinions. It is helpful to keep asking yourself: how strongly are my ideas shaped by my individual profiles of basic needs and competencies and my experiences, and how much are they influenced by the environment, or even forced on me against my will?

How do I react to Misfit situations?

You will find a detailed description of how to evaluate your emotional state and your mental and physical reactions to Misfit situations in Chapter 9. It can be difficult to assess just how much impact a Misfit situation has had on you, and how badly it has affected your mind and body. It can help to remember what you felt like during past Misfit situations, and why you found them upsetting. You could also ask the people around you – friends or colleagues, for example – how they see you.

The benchmark you should use for your assessment is the Fit situation, a condition in which you are largely in harmony with your environment. A Fit situation makes you feel at ease, with a good sense of self-worth and self-efficacy. If a Misfit situation arises, your well-being, self-worth and self-efficacy are the first things to be affected. If this situation gets worse, other symptoms will appear, depending on your individual mental and physical disposition. Answering some of the following questions will help you become more aware of your state of mind and the way you react to things:

- What are the indications that something has affected my wellbeing? What situations make me feel insecure? When do I feel helpless, and when am I overstretched?
- How do I react to these things? Am I anxious, irritable or gloomy?
- Have I become listless and less capable? Has my interest in work and hobbies waned?
- What is my relationship behaviour like? Am I withdrawn? Am I unusually preoccupied with myself?
- Am I plagued by physical symptoms such as headaches, tension, digestive problems or insomnia?
- Am I drinking more alcohol than I used to? Do I take medication more often?
- Do I feel emotionally empty? Am I suffering from depressive moods? Am I finding it difficult to cope with the challenges of everyday life?

A better understanding of your own basic needs, competencies and ideas, as well as your current state of mind and accompanying symptoms, can help you deal with a Misfit situation successfully.

Acknowledgements

My greatest thanks go to all the parents and children I got to know in the course of the Zurich Longitudinal Studies. They kept faith with the studies for decades and gave up their valuable time for them – something that was vital to the success of our research. From the children, I learned how diverse the path of child development can be as each child grows into a unique individual. The parents showed me what an admirable job they did every day in helping their children satisfy their basic needs. I would also like to thank the parents and children who placed themselves under my care in a clinical setting. From them, I learned how children can end up in Misfit situations, and what can be done to make things better for them.

Countless experiences with children and their parents have gone into the making of this book. I am also tremendously grateful to the academics, clinicians and teachers who enriched my knowledge of child development with their information and experiences and in so doing made fundamental contributions to the development of the Fit Principle. I feel I owe special thanks to my two teachers: Andrea Prader, the former Director of the University Children's Hospital in Zurich, and Arthur Hawley Parmelee, who was Director of the Child Development Unit at the University of California in Los Angeles. Huge thanks also go to all the colleagues in the Growth and Development Centre at the University Children's Hospital in Zurich, with whom I had the pleasure of working for more than thirty years. I would especially like to mention Caroline Benz, John Caflisch, Markus Schmid, Theo Gasser, Sepp Holtz, Oskar Jenni, Tanja Kakebeeke, Bea Latal, Luciano Molinari, Markus Schmid and Noa Stemmer-Holtz.

Heartfelt thanks to all the early readers, in particular Annina, Johanna and Kathrin Largo, who contributed greatly to the success of this book. I also received generous support from Monika Czernin, author and film maker, and Herbert Renz-Polster, author, paediatrician and academic. Their unrelenting emotional support was a constant source of strength for me and their constructive criticism always put me back on the right track. Heartfelt thanks also go to Alex Hajnal, developmental biologist, Albert Schinzel, geneticist, and Luciano Molinari, biostatistician, who shared their expert knowledge with me and saved me from errors.

This book would not have been written without the judicious support of Fischer Verlag. I would especially like to thank Nina Sillem, head of non-fiction, and her colleagues. Huge thanks to the editorial team, Alexander Roesler and Margret Trebbe-Plath. Margret supported me in the preparation of the manuscript with great wisdom and patience, showing real dedication and making a number of important suggestions. I would also like to thank Didier Ludwig, Peter Palm and Johanna Stierlin for their wonderful diagrams and drawings.

My warmest thanks go to my family, in particular to Brigitt and my daughters Eva, Kathrin and Johanna. They encouraged me in my project for more than six years and showed great interest in the book. I was always able to count on their patience and support, even during the difficult times that such a project involves.

Remo H. Largo
May 2017

Bibliography

30 000 Jahre Kunst: Künstlerisches Schaffen der Menschheit durch Zeit und Raum, Phaidon, 2015

Antonovsky, A., *Health, Stress, and Coping: New Perspectives on Mental and Physical Well-Being*, Jossey-Bass, 1979

Asendorpf, J. B., *Psychologie der Persönlichkeit*, Springer, 2005

Augustine, Saint, *The City of God*, trans. Marcus Dods, Hendrickson, 2009

Avery, O. T., C. M. MacLeod and M. McCarty, 'Studies on the chemical nature of the substance inducing transformation of pneumococcal types: Induction of transformation by a deoxyribonucleic acid fraction isolated from pneumococcus type III', *Journal of Experimental Medicine* 79, 2 (1944)

Baltes, P. B., and K. U. Mayer, *The Berlin Aging Study: Aging from 70 to 100*, Cambridge University Press, 2001

Bandura, A., *Social Learning Theory*, Prentice Hall, 1977

Bischof-Köhler, D., *Spiegelbild und Empathie: Die Anfänge der sozialen Kognition*, Huber, 1989

——, *Kinder auf Zeitreise: Theory of Mind, Zeitverständnis und Handlungsorganisation*, Huber, 2000

Blaffer Hrdy, S., *Mother Nature: Maternal Instincts and How They Shape the Human Species*, Ballantine Books, 2000

——, *Mothers and Others: The Evolutionary Origins of Mutual Understanding*, Belknap Press, 2009

Bowlby, J., *Attachment and Loss*, vol. 1: *Attachment*, New York, 1969; vol. 2: *Separation*, New York, 1975

——, *Frühe Bindung und kindliche Entwicklung*, Ernst Reinhardt, 2001

Brody, L. R., and J. A. Hall, 'Gender and emotion', in: M. Lewis and J. M. Haviland (eds.), *Handbook of Emotions*, Guilford Press, 1993, pp. 447–60

Buber, M., *Ich und Du*, Reclam, 2008

Caplan, N., *The Boat People and Achievement in America: A Study of Family Life, Hard Work, and Cultural Values*, University of Michigan Press, 1989

Chess, S., and A. Thomas, *Origins and Evolution of Behavior Disorders*, Bruner and Mazel, 1984

Chomsky, N., *Aspects of the Theory of Syntax*, MIT Press, 1967

Club of Rome, *The Limits to Growth: A Report for the Club of Rome's Project on the Predicament of Mankind*, Pan Books, 1974

Coradi Vellacott, M., and S. C. Wolter, *Chancengleichheit im schweizerischen Bildungswesen*, Aarau, 2005

Csíkszentmihályi, M., *Flow: The Psychology of Optimal Experience*, Harper and Row, 1990

Damasio, A., *The Feeling of What Happens: Body, Emotion and the Making of Consciousness*, Vintage, 2000

Darwin, C., *On the Origin of Species*, Harvard Classics, 1975

——, *The Descent of Man, and Selection in Relation to Sex*, Princeton University Press, 1981

——, Letter to Joseph D. Hooker, 29 March 1863, in *The Works of Charles Darwin*, vol. 19: *Variation of Animals and Plants under Domestication*, vol. I, Pickering, 1988, p. 10

Deary, I. J., *Intelligence: A Very Short Introduction*, Oxford University Press, 2001

Descartes, R., *Meditationes de prima philosophia*, Michael Soly, 1641

Deschner, K., *Kriminalgeschichte des Christentums* (10 vols.), Rowohlt, 2010

Diamond, A., 'Executive functions', *Annual Review of Psychology* 64 (2013), 135–68

Diamond, J., *Why is Sex Fun? The Evolution of Human Sexuality*, Basic Books, 1998

——, 'The worst mistake in the history of the human race', *Discover Magazine*, 1 May 1999

Dixon, S., et al., 'Early social interaction of infants with parents and strangers', *Journal of the American Academy of Child Psychiatry* 20:1 (1981), 32–52

Dornes, M., *Die Psychologie von René A. Spitz: Eine Einführung und kritische Würdigung*, Asanger Roland Verlag, 1981

Dunkake, I., et al., 'Schöne Schüler, schöne Noten? Eine empirische Untersuchung zum Einfluss der physischen Attraktivität von Schülern auf die Notenvergabe durch das Lehrpersonal', *Zeitschrift für Soziologie* 41:2 (2012), 142–61

Eaton, W. O., N. A. McKeen and D. W. Campbell, 'The waxing and waning of movement: Implications for psychological development', *Developmental Review* 21:2 (2001), 205–23

Eccles, J. C., *Mind and Brain*, Paragon House, 1999

Eibl-Eibesfeldt, I., *Grundriß der vergleichenden Verhaltensforschung*, Piper, 1974

——, *Die Biologie des menschlichen Verhaltens: Grundriss der Humanethologie*, Blank Media, 2004

Eimas, P. D., et al., 'Speech perception in infants', *Science* 171:3968 (1971), 303–6

Elsberg, M., *Blackout: Tomorrow Will be Too Late*, Black Swan, 2017

Eveleth, B., and J. M. Tanner, *Worldwide Variation in Human Growth*, Cambridge University Press, 1976

Falkner, F., and J. M. Tanner (eds.), *Human Growth*, vol. 2, Plenum Press, 1986

Fantz, R. L., 'Visual perception from birth as shown by pattern selectivity', *Annals of New York Academic Science* 118:21 (1965), 793–814

Fischer, L., 'DNA-Reparatur gegen Krebs und Altern', *Spektrum der Wissenschaft*, 19 November 2015

Flavell, J. H., F. L. Green, and E. R. Flavell, 'Children's understanding of the stream of consciousness', *Child Development* 64:2 (1993), 387–98

Flavell, J. H., et al., 'The development of children's knowledge about inner speech', *Child Development* 68:1 (1997), 39–47

Flynn, J. R., 'The mean IQ of Americans: Massive gains 1932 to 1978', *Psychological Bulletin* 95:1 (1984), 29–51

——, 'Massive IQ gains in 14 nations: What IQ tests really measure', *Psychological Bulletin* 101:2 (1987), 171–91

Fölsing, A., *Albert Einstein*, Penguin Books, 1997

Forrest, D. W., *Francis Galton: The Life and Work of a Victorian Genius*, Elek Books, 1974

Futuyma, D. J., *Evolutionary Biology*, 3rd rev. edn, Sinauer Associates, 1986

Fux, B., et al.: 'Inequality in Switzerland', *Swiss Journal of Sociology* 28:2 (2002)

Gallahue, D. L., *Understanding Motor Development: Infants, Children, Adolescents*, McGraw-Hill, 1998

Gallup, G. G., 'Self-recognition in primates: A comparative approach to the bidirectional properties of consciousness', *American Psychologist* 32:5 (1977), 329–38

Gardner, H., *Frames of Mind: The Theory of Multiple Intelligences*, Basic Books, 1985

——, *Multiple Intelligences: The Theory in Practice, a Reader*, Basic Books, 1993

Gegenbaur, C., *Grundzüge der vergleichenden Anatomie*, Engelmann, 1870

Gendlin, E. T., *Focusing: How to Gain Direct Access to Your Body's Knowledge*, Bantam, 1982

Gigerenzer, G., *Gut Feelings: Short Cuts to Better Decision Making*, Penguin Books, 2008

Gilligan, C., *In a Different Voice: Psychological Theory and Women's Development*, Harvard University Press, 1982

Goleman, D., *Emotional Intelligence: Why It Can Matter More Than IQ*, Bloomsbury, 1996

Goodall, J., and P. Berman, *Reason for Hope: A Spiritual Journey*, Thorsons, 2000

Green, R. E., et al., 'A draft sequence of the Neanderthal Genome', *Science* 328:5979 (2010), 710–22

Griffin, A. M., and J. H. Langlois, 'Stereotype directionality and attractiveness stereotyping: Is beauty good or is ugly bad?', *Social Cognition*, 24:2 (2006), 187–206

Griffiths, J. F., et al., *Introduction to Genetic Analysis*, W. H. Freeman, 2012

Grüter, T., and M. Grüter, 'Last but not least: Prosopagnosia in biographies and autobiographies', *Perception* 36:2 (2007), 299–301

Haines, D. W. (ed.), *Refugees as Immigrants: Cambodians, Laotians and Vietnamese in America*, Rowman & Littlefield, 1989

Harari, Y., *Sapiens: A Brief History of Humankind*, Harvill Secker, 2014

Harris, J. R., *The Nurture Assumption: Why Children Turn Out the Way They Do*, Free Press, 1998

Hattie, J., *Visible Learning: A Synthesis of over 800 Meta-Analyses Relating to Achievement*, Routledge, 2009

——, *Visible Learning for Teachers: Maximizing Impact on Learning*, Routledge, 2012

Hayes, B., 'Die neuronalen Netzwerke werden erwachsen', *Spektrum der Wissenschaft*, 15 August 2014

Hebb, D. O., *The Organization of Behavior: A Neuropsychological Theory*, Erlbaum Books, 1949

Herrero, J., 'European Molecular Biology Laboratory, European Bioinformation Institute', *National Geographic*, July 2013, p. 102

Hobaiter, C., and R. W. Byrne, 'The gestural repertoire of the wild chimpanzee', *Animal Cognition* 14:5 (2011), 745–67

Höffe, O., *Gerechtigkeit: Eine philosophische Einführung*, Beck, 2004

Hubel, D. H., et al., 'Auge und Gehirn: Neurobiologie des Sehens', *Spektrum der Wissenschaft*, 1995

Hurrelmann, K., and O. Razum (eds.), *Handbuch der Gesundheitswissenschaften*, Juventa, 2006

Huxley, T. H., *Man's Place in Nature*, D. Appelton, 1863

Iglowstein, I., et al., 'Sleep duration from infancy to adolescence: Reference values and generational trends', *Pediatrics* 111:2 (2003), 302–7

Jablonski, N. G., and G. Chaplin, 'Human skin pigmentation as an adaptation to UV radiation', *PNAS* 107 (2010), Supplement 2, 8962–8

Jensen, A. R., and L. J. Wang, 'What is a good g?', *Intelligence* 18:3 (1994), 231–58

Kanaya, T., M. H. Scullin and S. J. Ceci, 'The Flynn effect and U.S. policies: The impact of rising IQ scores on American society via mental retardation diagnosis', *American Psychologist* 58:10 (2003), 778–90

Kegel, B., *Epigenetik: Wie unsere Erfahrungen vererbt werden*, Dumont, 2015

Keller, H., Y. H. Poortinga and A. Schölmerich (eds.), *Between Culture and Biology*, Cambridge University Press, 2002

Kohlberg, L., 'Moral stage and moralization: The cognitive-developmental approach', in T. Lickona (ed.), *Moral Development and Behavior: Theory, Research and Social Issues*, Holt, Rinehart & Winston, 1976

——, *The Psychology of Moral Development*, Harper & Row, 1984

Krause, J., et al., 'The complete mitochondrial DNS genome of an unknown hominin from southern Siberia', *Nature* 464:7290 (2010), 894–7

Kronig, W., *Die systematische Zufälligkeit des Bildungserfolgs: Theoretische Erklärungen und empirische Untersuchungen und Leistungsbewertung von leistungsschwachem Lernen*, Haupt, 2007

Lander, E. S., et al., 'Human genome: Initial sequencing and analysis of the human genome', *Nature* 409:6822 (2001), 860–921

Largo, R. H., 'Catch-up growth during adolescence', *Hormone Research* 39 (1993), Supplement 3, 41–8

——, *Kinderjahre: Die Individualität des Kindes als erzieherische Herausforderung*, Piper, 1999

——, *Babyjahre: Entwicklung und Erziehung in den ersten vier Jahren*, Piper, 2007

——, *Lernen geht anders: Bildung und Erziehung vom Kind her denken*, Piper, 2012

——, *Wer bestimmt den Schulerfolg: Kind, Schule, Gesellschaft?* Beltz, 2013

—— and M. Beglinger, *Schülerjahre: Wie Kinder besser lernen*, Piper, 2009

—— and M. Czernin, *Jugendjahre: Kinder durch die Pubertät begleiten*, Piper, 2011

—— and ——, *Glückliche Scheidungskinder: Was Kinder nach der Trennung brauchen*, Piper, 2014

—— and J. A. Howard, 'Developmental progression in play behavior of children between nine and thirty months of age: I. Spontaneous play and imitation', *Developmental Medicine and Child Neurology* 21:3 (1979), 299–310

—— and ——, 'Developmental progression in play behavior of children between nine and thirty months of age: II. Spontaneous play and language development' in: *Developmental Medicine and Child Neurology*, 21:4 (1979), 492–503

——, J. A. Caflisch, et al., 'Neuromotor development from 5 to 18 years: Part 1. Timed performance', *Developmental Medicine and Child Neurology* 43:7 (2001), 436–43

——, L. Comenale Pinto, et al., 'Language development during the first five years of life in term and preterm children: Significance of pre-, peri- and postnatal events', *Developmental Medicine and Child Neurology* 28:3 (1986), 333–50

——, L. Molinari, et al., 'Early development of locomotion: Significance of prematurity, cerebral palsy and sex', *Developmental Medicine and Child Neurology* 27:2 (1985), 183–91

Leakey, R. E., *The Origin of Humankind*, Perseus Books, 1994

Lenneberg, E. H., *Biological Foundations of Language*, John Wiley and Sons, 1967

Levy, R., et al., *Tous égaux? De la stratification aux représentations*. Editions Seismo, 1997

Libet, B. W., 'Do we have a free will?', *Journal of Consciousness Studies* 6: 8–9 (1999), 47–57

Lisker L., and A. S. Abramson, *The Voicing Dimensions: Some Experiments in Comparative Phonetics*, Academia, 1970

Lorenz, K., *Vergleichende Verhaltensforschung, oder, Grundlagen der Ethologie*, Springer, 1978

Mai, K. R., *Die Bachs: Eine deutsche Familie*, Propyläen, 2013

Mampel, B., et al., 'Newborns' cry melody is shaped by their native language', *Current Biology* 19:23 (2009), 1994–7

Marshall, A. M., and J. M. Tanner, 'Puberty', in F. Falkner and J. M. Tanner (eds.), *Human Growth: A Comprehensive Treatise*, vol. 2: *Developmental Biology Prenatal Growth*, Plenum Press, 1986

Maslow, A. H., *Motivation and Personality*, Pearson, 1997

Maturana, H. R., and F. Varela, *The Tree of Knowledge: The Biological Roots of Human Understanding*, Shambhala, 1992

Mayr, E., *Das ist Evolution*, Bertelsmann, 2003

McKone, E., K. Crookes and N. Kanwisher, 'The cognitive and neural development of face recognition in humans', in M. Gazzaniga (ed.), *The Cognitive Neurosciences* (4th edn), MIT Press, 2009, pp. 467–82

Melzoff, A., and M. K. Moore, 'Imitations of facial and manual gestures by human neonates', *Science* 198:4312 (1977), 75–8

Mendel, G., 'Versuche über Pflanzen-Hybriden', in *Verhandlungen des Naturforschenden Vereines in Brünn*, vol. IV (Abhandlungen 1865), Brünn, 1866, pp. 3–47

Merimee, T. J., et al., 'Insulin-like growth factors in pygmies: The role of puberty in determining final stature', *New England Journal of Medicine* 316:15 (1987), 906–11

Moffat, A., and J. Wilson, *The Scots: A Genetic Journey*, Birlinn, 2011

Molcho, S., *Body Speech*, Sun Books, 1985

Möllers, N., C. Schwägerl and H. Trischler, *Willkommen im Anthropozän: Unsere Verantwortung für die Zukunft der Erde*, Deutsches Museum Verlag, 2015

Morris, D., *Bodywatching: A Field Guide to the Human Species*, Grafton, 1987

Moser, U., *Analyse zur Volksschule zuhanden der SP Schweiz*, Zurich, 2007

—— and F. Keller, *Check 5: Schlussbericht zuhanden des Departements Bildung, Kultur und Sport des Kantons Aargau*, Zurich, 2008

—— and S. Tresch, *Best Practice in der Schule: Von erfolgreichen Lehrerinnen und Lehrern lernen*, Zurich, 2003

——, N. A. Fox and C. H. Zeanah, 'Die entscheidenden zwei Jahre', *Spektrum der Wissenschaft*, 17 January 2014

Nelson, C., C. H. Zeanah, et al., 'Cognitive recovery in socially deprived young children: The Bucharest early intervention project', *Science* 318:5858 (2007), 1937–40

Nesselrode, J. R., and R. B. Cattel, *Handbook of Multivariate Experimental Psychology*, Rand McNally, 1966

Neubauer, A., and E. Stern, *Lernen macht intelligent: Warum Begabung gefördert werden muss*, DVA, 2007

Nowicki, S., and M. P. Duke, 'Individual differences in the nonverbal communication of affect: The diagnostic analysis of nonverbal accuracy scale', *Journal of Nonverbal Behavior* 18:1 (1994), 9–35

OECD, 'Bildung auf einen Blick', September 2011: http://www.oecd.org/berlin/publikationen/bildungaufeinenblick2011.htm

OECD, PISA Study, 2009: http://www.pisa.oecd.org

Olson, S., *Mapping Human History: Discovering the Past through Our Genes*, Bloomsbury, 2002

Pan, Y., and W. Ke-Sheng, 'Spousal concordance in academic achievements and IQ. A principal component analysis', *Open Journal of Psychiatry* 1:02 (2011), 15–19

Papousek, H., and M. Papousek, 'Early ontogeny of human social interaction: Its biological roots and social dimensions', in M. V. Cranach et al. (eds.), *Human Ethology: Claims and Limits of a New Discipline*, Cambridge University Press, 1979

—— and ——, 'Lernen im ersten Lebensjahr', in L. Montada (ed.), *Brennpunkte der Entwicklungspsychologie*, Kohlhammer, 1989

Pauen, S., 'Zeitfenster der Gehirn- und Verhaltensentwicklung: Modethema oder Klassiker?', in H. Herrmann (ed.), *Neurodidaktik: Grundlagen und Vorschläge für gehirngerechtes Lehren und Lernen*, Beltz, 2006

Perner, J., and H. Wimmer, 'John thinks that Mary thinks that: Attribution of second order beliefs by 5- to 10-year-old children', *Journal of Experimental Psychology* 39:3 (1985), 437–71

Piaget, J., *The Origin of Intelligence in the Child*, Routledge, 1997

——, *The Language and Thought of the Child: Selected Works*, vol. 5, trans. Marjorie and Ruth Gabain, Routledge, 2014

Pike, A. W. G., et al., 'U-series dating of paleolithic art in 11 caves in Spain', *Science* 33, 2012

Piketty, T., *Capital in the Twenty-First Century*, Belknap Press, 2014

Pinker, S., *The Better Angels of Our Nature: A History of Violence and Humanity*, Penguin Books, 2011

——, *The Village Effect: Why Face-to-Face Contact Matters*, Atlantic Books, 2015

Plomin, R., *Nature and Nurture: An Introduction to Human Behavioral Genetics*, Brooks/Cole Pacific Grove, 1990

Pollard, K. S., 'Der feine Unterschied', *Spektrum der Wissenschaft*, 26 June 2009

Popper, K., *The Logic of Scientific Discovery*, Routledge, 2002

Portmann, A., 'Die biologische Bedeutung des ersten Lebensjahres beim Menschen', *Schweizerische Medizin: Wochenzeitschrift* 71 (1941), 921–1001

Postman, N., *Amusing Ourselves to Death: Public Discourse in the Age of Show Business*, Penguin Books, 1986

Prechtl, H. F. R., and D. Beintema, *The Neurological Examination of the Full Term Newborn Infant*, Heinemann, 1964

Premack, D., and A. Premack, *The Mind of an Ape*, W. W. Norton, 1983

Premack, D., and G. Woodruff, 'Does the chimpanzee have a theory of mind?', *Behavioral Brain Science* 1:4 (1978), 515–26

Ramsey, J. L., and J. H. Langlois, 'Effects of the "beauty is good" stereotype on children's information processing', *Journal of Experimental Child Psychology* 81:3 (2002), 320–40

Rawls, J., *A Theory of Justice*, Harvard University Press, 1971

Reich, D., et al., 'Genetic history of an archaic hominin group from Denisova Cave in Siberia', *Nature* 468:7327 (2010), 1053–60

Renz-Polster, H., *Kinder verstehen. Born to be wild: Wie die Evolution unsere Kinder prägt*, Kösel, 2014

——, *Menschenkinder: Artgerechte Erziehung – was unser Nachwuchs wirklich braucht*, Kösel, 2016

Rizzolatti, G., and C. Sinigaglia, *Mirrors in the Brain: How Our Minds Share Actions and Emotions*, Oxford University Press, 2008

Röhrlich, D., *Urmeer: Die Entstehung des Lebens*, Mare Verlag, 2012

Rölleke, H., *Kinder- und Hausmärchen: Gesammelt durch die Gebrüder Grimm*, Wissenschaftliche Buchgesellschaft, 1999

Sameroff, A. J., and M. J. Chandler, 'Reproductive risk and the continuum of caretaking casualty', in F. D. Horowitz, et al. (eds.), *Review of Child Development Research* 4, University of Chicago Press, 1975

Scarr, S., 'Developmental theories for the 1990s: Development and individual differences', *Child Development* 63:1 (1992), 1–19

Schneider, W., and U. Lindenberger (formerly Oerter and Montada), *Entwicklungspsychologie*, Beltz, 2012

Schulte-Markwort, M., *Burnout Kids: Wie das Prinzip Leistung unsere Kinder überfordert*, Pattloch, 2016

Sciencemag.org, 'Tibetans inherited high-altitude gene from ancient human', 4 July 2014

Seksik, L., *Der Fall Eduard Einstein*, Blessing, 2014

Sennett, R., *The Craftsman*, Yale University Press, 2008

Simonton, D. K., *Creativity in Science: Chance, Logic, Genius and Zeitgeist*, Cambridge University Press, 2004

Singer, W., 'Neuronal synchrony: A versatile code for the definition of relations?', *Neuron* 24:1 (1999), 49–65

——, *Der Beobachter im Gehirn: Essays zur Hirnforschung*, Suhrkamp, 2002

——, 'Synchrony, oscillations, and relational codes', in L. M. Chalupal and J. S. Werner, *The Visual Neurosciences*, MIT Press, 2004

Slomkowski, C., and J. Dunn, 'Young children's understanding of other people's beliefs and feelings and their connected communication with friends', *Developmental Psychology* 32 (1996)

Sodian, B., 'The development of deception in young children', *British Journal of Developmental Psychology* 9:1 (1991), 173–88

Sommer, M., *Evolutionäre Anthropologie zur Einführung*, Junius, 2015

Starkey, P., E. S. Selke and R. Gelman, 'Numerical abstraction by human infants', *Cognition* 36:2 (1990), 97–127

Staub, K., *Der biologische Lebensstandard in der Schweiz seit 1800: Historisch-anthropometrische Untersuchung der Körperhöhe (und des Körpergewichts) in der Schweiz seit 1800, differenziert nach Geschlecht, sozioökonomischem und regionalem Hintergrund*, doctoral thesis, University of Bern, 2010

——, et al., 'The average height of 18- and 19-year-old conscripts (N = 458,322) in Switzerland from 1992 to 2009, and the secular height trend since 1878', *Swiss Medical Weekly* 141, 30 July 2011

Steinzeitmusik: http://www.spektrum.de/news/steinzeitmusik/999398

Straubhaar, T., *Radikal gerecht: Wie das bedingungslose Grundeinkommen den Sozialstaat revolutioniert*, Edition Körber-Stiftung, 2017

Sundet, J. M., D. G. Barlaug and T. M. Torjussen, 'The end of the Flynn effect? A study of secular trends in mean intelligence test scores of Norwegian conscripts during half a century', *Intelligence* 32 (2004), 349–62

Szagun, G., *Sprachentwicklung beim Kind*, Beltz, 2006

Tammet, D., *Thinking in Numbers: How Maths Illuminates Our Lives*, Hodder & Stoughton, 2012

Teasdale, T. W., and D. R. Owen, 'A long-term rise and recent decline in intelligence test performance: The Flynn effect in reverse', *Personality and Individual Differences* 39:4 (2005), 837–43

Tomasello, M., *The Cultural Origins of Human Cognition*, Harvard University Press, 2001

——, *Origins of Human Communication*, MIT Press, 2008

——, *Why We Cooperate*, MIT Press, 2009

Ulijaszek, S. J., F. E. Johnston and M. A. Preece, *The Cambridge Encyclopedia of Human Growth and Development*, Cambridge University Press, 1998

Van Wieringen, J. C., 'Secular growth changes', in F. Falkner and J. M. Tanner (eds.), *Human Growth*, vol. 3, 1986, pp. 307–32

Varela, F. J., H. R. Maturana and R. Uribe, 'Autopoiesis: The organization of living systems, its characterization and a model', *Biosystems,* 5:4 (1974), 187–96

Watson, J. S., 'Smiling, cooing and "the game"', *Merrill-Palmer Quarterly of Behavior and Development* 18:4 (1972), 323–39

Watzlawick, P., J. H. Beavin and D. D. Jackson, *Pragmatics of Human Communication: A Study of Interaction Patterns, Pathologies and Paradoxes,* W. W. Norton, 2011

Weiner, J., *Der Schnabel des Finken, oder, Der kurze Atem der Evolution,* Knaur, 1994

Werner, E. E., *The Children of Kauai: A Longitudinal Study from the Prenatal Period to Age Ten,* University of Hawaii Press, 1971

——, *Vulnerable But Invincible: A Longitudinal Study of Resilient Children and Youth,* McGraw-Hill, 1989

WHO: http://www.who.int/childgrowth/standards/en

Wilson, R. S., 'The Louisville Twin Study: Developmental synchronies in behavior', *Child Development* 54:2 (1983), 298–316

Wimmer, H., and J. Perner, 'Beliefs about beliefs: Representation and constraining function of wrong belief in young children's understanding of deception', *Cognition* 13:1 (1983), 103–28

Winfree, A. T., *The Geometry of Biological Time,* Springer, 1980

World Values Survey 2001: http://www.worldvaluessurvey.org

Zalasiewicz, J., et al., 'Are we now living in the Anthropocene?', *GSA Today* 18:2 (2008), 4–8

Zaroff, C. M., and R. C. D'Amato, *The Neuropsychology of Men: A Developmental Perspective from Theory to Evidence-Based Practice,* Springer, 2015

Notes

EPIGRAPH

1. From a letter written in September 1931. Hermann Hesse, *Lektüre für Minuten: Gedanken aus seinen Büchern und Briefen* (ed. Volker Michels), Suhrkamp Verlag, 1971, p. 90. Translated by Ruth Martin.

INTRODUCTION

1. Pindar, *Pythian Odes* 2, line 72.

CHAPTER 1: HUMAN BIOLOGICAL AND SOCIO-CULTURAL DEVELOPMENT

1. Darwin, *On the Origin of Species*, p. 529.
2. Darwin, letter to Joseph D. Hooker.
3. Mayr, *Das ist Evolution*.
4. Röhrlich, *Urmeer*.
5. Mendel, 'Versuche über Pflanzen-Hybriden'.
6. Mayr, *Das ist Evolution*.
7. Fischer, 'DNA-Reparatur gegen Krebs und Altern'.
8. Ibid.
9. Kegel, *Epigenetik*.
10. For an in-depth look at this, see Diamond, *Why is Sex Fun?*; Morris, *Bodywatching*.
11. Lander et al., 'Human genome'.
12. Griffiths et al., *Introduction to Genetic Analysis*.
13. Moffat and Wilson, *The Scots*.
14. Olson, *Mapping Human History*.

15. Jablonski and Chaplin, 'Human skin pigmentation as an adaptation to UV radiation'.
16. Eveleth and Tanner, *Worldwide Variation in Human Growth*.
17. Ibid.
18. Merimee et al., 'Insulin-like growth factors in pygmies'.
19. Asendorpf, *Psychologie der Persönlichkeit*.
20. Herrero, 'European Molecular Biology Laboratory, European Bioinformation Institute'.
21. Leakey, *The Origin of Humankind*.
22. Krause et al., 'The complete mitochondrial DNS genome of an unknown hominin from southern Siberia'; Reich et al., 'Genetic history of an archaic hominin group from Denisova Cave in Siberia'.
23. Green et al., 'A draft sequence of the Neanderthal Genome'.
24. Sciencemag.org, 'Tibetans inherited high-altitude gene from ancient human'.
25. Pollard, 'Der feine Unterschied'.
26. Prechtl and Beintema, *The Neurological Examination of the Full Term Newborn Infant*.
27. Portmann, 'Die biologische Bedeutung des ersten Lebensjahres beim Menschen'.
28. Olson, *Mapping Human History*.
29. Ibid.
30. *30 000 Jahre Kunst*.
31. Diamond, 'The worst mistake in the history of the human race'.
32. Tomasello, *The Cultural Origins of Human Cognition*.
33. Harari, *Sapiens*.
34. Blaffer Hrdy, *Mothers and Others*.
35. Pinker, *The Better Angels of Our Nature*.
36. Zalasiewicz et al., 'Are we now living in the Anthropocene?'
37. In Albert Baskerville (ed.), *The Poetry of Germany*, New York 1854, pp. 7–8.
38. Ernst, Freiherr von Feuchtersleben, *Beiträge zur Literatur, Kunst-, und Lebenstheorie*, vol. 2, III: *Blätter aus dem Tagebuche des Einsamen*, Verlag von Josef Stockhölzer u. a., 1841.

CHAPTER 2: THE COMBINED EFFECT OF GENETIC PREDISPOSITION AND ENVIRONMENT

1. Schneider and Lindenberger, *Entwicklungspsychologie*.
2. Sameroff and Chandler, 'Reproductive risk and the continuum of care-taking casualty'.
3. Forrest, *Francis Galton*; Simonton, *Creativity in Science*.
4. Pan and Ke-Sheng, 'Spousal concordance in academic achievements and IQ'.
5. Fölsing, *Albert Einstein*.
6. Seksik, *Der Fall Eduard Einstein*.
7. Mai, *Die Bachs*.
8. Staub, *Der biologische Lebensstandard in der Schweiz seit 1800*.
9. Van Wieringen, 'Secular growth changes'.
10. Eveleth and Tanner, *Worldwide Variation in Human Growth*.
11. Staub et al., 'The average height of 18- and 19-year-old conscripts'.
12. Marshall and Tanner, 'Puberty'.
13. Flynn, 'The mean IQ of Americans'; Flynn, 'Massive IQ gains in 14 nations'.
14. Sundet et al., 'The end of the Flynn effect?'.
15. Kanaya et al., 'The Flynn effect and U.S. policies'.
16. Teasdale and Owen, 'A long-term rise and recent decline in intelligence test performance'.
17. Sundet et al., 'The end of the Flynn effect?'.
18. OECD, PISA Study.
19. Wilson, 'The Louisville Twin Study'; Plomin, *Nature and Nurture*; Scarr, 'Developmental theories for the 1990s'.
20. Wilson, 'The Louisville Twin Study'.
21. Scarr, 'Developmental theories for the 1990s'.
22. Ibid.
23. Ibid.
24. Harris, *The Nurture Assumption*.
25. Largo, Comenale Pinto, et al., 'Language development during the first five years of life in term and preterm children'.
26. Largo, 'Catch-up growth during adolescence'.
27. Levy et al., *Tous égaux?*
28. Coradi Vellacott and Wolter, *Chancengleichheit im schweizerischen Bildungswesen*; Kronig, *Die systematische Zufälligkeit des Bildungserfolgs*.
29. *30 000 Jahre Kunst*.
30. Schulte-Markwort, *Burnout Kids*.

CHAPTER 3: DEVELOPING INTO INDIVIDUALS

1. Ambrose Bierce, *The Devil's Dictionary*, Neale Publishing, 1911.
2. Zaroff and D'Amato, *The Neuropsychology of Men*.
3. Hebb, *The Organization of Behavior*.
4. Eccles, *Mind and Brain*.
5. Fantz, 'Visual perception from birth as shown by pattern selectivity'.
6. McKone et al., 'The cognitive and neural development of face recognition in humans'.
7. Hayes, 'Die neuronalen Netzwerke werden erwachsen'.
8. Hubel et al., 'Auge und Gehirn'.
9. Singer, 'Neuronal synchrony'; Singer, 'Synchrony, oscillations, and relational codes'.
10. Watson, 'Smiling, cooing and "the game"'.
11. Csíkszentmihályi, *Flow*.
12. Neubauer and Stern, *Lernen macht intelligent*.
13. Dornes, *Die Psychologie von René A. Spitz*.
14. Nelson et al., 'Cognitive recovery in socially deprived young children'; Moser et al., 'Die entscheidenden zwei Jahre'.
15. Baltes and Mayer, *The Berlin Aging Study*.
16. Gardner, *Frames of Mind*.
17. Nesselrode and Cattell, *Handbook of Multivariate Experimental Psychology*.
18. Ibid.

CHAPTER 4: BASIC NEEDS THAT SHAPE OUR LIVES

1. Jean-Jacques Rousseau, *Emile*, book 4, 'The Creed of a Savoyard Priest'.
2. http://www.biogetica.com/i_am_that.pdf.

CHAPTER 5: COMPETENCIES WE WANT TO DEVELOP

1. Georg Christoph Lichtenberg, *Philosophical Writings*, 'Notebook J', aphorism 1551: https://issuu.com/bouvard6/docs/lichtenberg_-_philosophical_writing, p. 139.
2. Jensen and Wang, 'What is a good g?'
3. Diamond, 'Executive functions'.

4. Gardner, *Multiple Intelligences*.
5. Watzlawick et al., *Pragmatics of Human Communication*.
6. Goleman, *Emotional Intelligence*.
7. Molcho, *Body Speech*; Morris, *Bodywatching*.
8. Eibl-Eibesfeldt, *Grundriss der vergleichenden Verhaltensforschung*.
9. Grüter and Grüter, 'Last but not least'.
10. Goodall and Berman, *Reason for Hope*.
11. Brody and Hall, 'Gender and emotion'.
12. Lorenz, *Vergleichende Verhaltensforschung*.
13. Dixon et al., 'Early social interaction of infants with parents and strangers'.
14. Bowlby, *Attachment and Loss*, vol. 1: *Attachment*; vol. 2: *Separation*.
15. Ibid.
16. Bandura, *Social Learning Theory*.
17. Mayr, *Das ist Evolution*.
18. Rizzolatti and Sinigaglia, *Mirrors in the Brain*.
19. Melzoff and Moore, 'Imitations of facial and manual gestures by human neonates'.
20. Largo and Howard, 'Developmental progression in play behavior of children between nine and thirty months of age: I.'
21. Piaget, *The Language and Thought of the Child*.
22. Confucius, *Analects of Confucius* 15.23.
23. *Mahabharata*, Anusasana Parva 113, 8.
24. Bischof-Köhler, *Spiegelbild und Empathie*.
25. Gallup, 'Self-recognition in primates'.
26. Premack and Woodruff, 'Does the chimpanzee have a theory of mind?'; Wimmer and Perner, 'Beliefs about beliefs'; Bischof-Köhler, *Spiegelbild und Empathie*.
27. Flavell et al., 'Children's understanding of the stream of consciousness'; Flavell et al., 'The development of children's knowledge about inner speech'.
28. Sodian, 'The development of deception in young children'.
29. Perner and Wimmer, 'John thinks that Mary thinks that'; Slomkowski and Dunn, 'Young children's understanding of other people's beliefs and feelings'.
30. Keller et al., *Between Culture and Biology*.
31. Kohlberg, 'Moral stage and moralization'; Kohlberg, *The Psychology of Moral Development*.
32. Gilligan, *In a Different Voice*.
33. Rawls, *A Theory of Justice*.

34. Conrad Ferdinand Meyer, 'Roman Fountain', trans. A. Z. Foreman, http://poemsintranslation.blogspot.co.uk/2009/06/conrad-ferdinand-meyer-roman-fountain.html.
35. Premack and Premack, *The Mind of an Ape*.
36. Tomasello, *Why We Cooperate*.
37. Hobaiter and Byrne, 'The gestural repertoire of the wild chimpanzee'.
38. Chomsky, *Aspects of the Theory of Syntax*.
39. Eimas et al., 'Speech perception in infants'.
40. Lisker and Abramson, *The Voicing Dimensions*.
41. Largo and Howard, 'Developmental progression in play behavior of children between nine and thirty months of age: II'.
42. Largo, Comenale Pinto, et al., 'Language development during the first five years of life in term and preterm children; Szagun, *Sprachentwicklung beim Kind*.
43. Lenneberg, *Biological Foundations of Language*.
44. Victor Hugo, *William Shakespeare*, part I, book II, chapter IV.
45. Mampel et al., 'Newborns' cry melody is shaped by their native language'.
46. Papousek and Papousek, 'Early ontogeny of human social interaction'.
47. Fantz, 'Visual perception from birth as shown by pattern selectivity'.
48. Popper, *The Logic of Scientific Discovery*.
49. Starkey et al., 'Numerical abstraction by human infants'.
50. Piaget, *The Origin of Intelligence in the Child*.
51. St Augustine, *The Confessions of Saint Augustine*, book XI, 14, trans. Edward Bouverie Pusey (1909–14): http://www.sacred-texts.com/chr/augconf/aug11.htm.
52. Winfree, *The Geometry of Biological Time*.
53. Largo, Molinari, et al., 'Early development of locomotion'; Gallahue, *Understanding Motor Development*.
54. Dunkake et al., 'Schöne Schüler, schöne Noten?'
55. Griffin and Langlois, 'Stereotype directionality and attractiveness stereotyping'.
56. Ramsey and Langlois, 'Effects of the "beauty is good" stereotype on children's information processing'.
57. Tammet, *Thinking in Numbers*.
58. Largo and Beglinger, *Schülerjahre*; Largo and Czernin, *Jugendjahre*.
59. Sennett, *The Craftsman*.

CHAPTER 6: OUR IDEAS AND BELIEFS

1. Friedrich Nietzsche, *Twilight of the Idols*, 'Maxims and Arrows' (1889).
2. Largo and Howard, 'Developmental progression in play behavior of children between nine and thirty months of age: I'.
3. Largo and Howard, 'Developmental progression in play behavior of children between nine and thirty months of age: II'.
4. Höffe, *Gerechtigkeit*.
5. Descartes, *Meditationes de prima philosophia*.
6. Damasio, *The Feeling of What Happens*.
7. Gigerenzer, *Gut Feelings*.
8. Libet, 'Do we have a free will?'
9. Michel de Montaigne, *Essays* III.
10. From a speech to the Royal College of Surgeons, 1923.

CHAPTER 7: FROM NATURE TO THE MAN-MADE ENVIRONMENT

1. Johann Wolfgang von Goethe, *Scientific Writings* (*Goethes Werke, Hamburger Ausgabe*, vol. 13, p. 17.
2. Möllers et al., *Willkommen im Anthropozän*.
3. Renz-Polster, *Kinder verstehen*.
4. Pinker, *The Better Angels of Our Nature*.
5. OECD.
6. Postman, *Amusing Ourselves to Death*.
7. Tomasello, *Why We Cooperate*.
8. Bowlby, *Frühe Bindung und kindliche Entwicklung*.
9. Schulte-Markwort, *Burnout Kids*.
10. Pinker, *The Village Effect*.
11. John Steinbeck, *Of Mice and Men* [1937], Penguin Books, 2006.
12. Friedrich Dürrenmatt, from the essay: 'Switzerland: A Prison' – a speech for Vaclav Havel, 1990.

CHAPTER 8: LIVING THE RIGHT LIFE: THE FIT PRINCIPLE

1. Buber, *Ich und Du*.
2. Maslow, *Motivation and Personality*.
3. Varela et al., 'Autopoiesis'; Maturana and Varela, *The Tree of Knowledge*.

4. Antonovsky, *Health, Stress, and Coping*; Hurrelmann and Razum, *Handbuch der Gesundheitswissenschaften*.
5. Chess and Thomas, *Origins and Evolution of Behavior Disorders*.
6. Iglowstein et al., 'Sleep duration from infancy to adolescence'.
7. Largo and Czernin, *Jugendjahre*; Largo and Czernin, *Glückliche Scheidungskinder*.
8. Largo, *Kinderjahre*.
9. Largo and Beglinger, *Schülerjahre*; Largo, *Lernen geht anders*; Largo, *Wer bestimmt den Schulerfolg*.
10. Largo and Czernin, *Jugendjahre*.
11. Largo and Czernin, *Glückliche Scheidungskinder*.
12. Johann Gottlieb Fichte, *Early Philosophical Writings*, trans. Daniel Breazeale, Cornell University Press, 1988, p. 159.

CHAPTER 9: MISFIT CONSTELLATIONS

1. Lao Tzu, quoted in Carl Rogers, *A Way of Being*, Houghton Mifflin, 1980, p. 42.
2. Werner, *The Children of Kauai*; Werner, *Vulnerable But Invincible*.
3. Caplan, *The Boat People and Achievement in America*; Haines, *Refugees as Immigrants*.
4. Gendlin, *Focusing*.
5. Thomas Merton, from a speech, quoted in Michael W. Fox, *Religious Education*, vol. 73 (1978), p. 292.

CHAPTER 10: CHANGING TIMES

1. Elsberg, *Blackout*.
2. http://docplayer.net/23992056-In-america-manufacturing-themythand-the-reality-of-michaelj-hicks-phd-srikant-devaraj-msmba-pmp-june-2015-ball-stateuniversity.html.
3. http://www.oxfordmartin.ox.ac.uk/downloads/academic/The_Future_of_Employment.pdf.
4. http://www.mckinsey.com/business-functions/digitalmckinsey/our-insights/disruptive-technologies.
5. Straubhaar, *Radikal gerecht*.
6. Global Wealth Report: https://www.allianz.com/v_1411376188000/media/economic_research/publications/specials/de/AGWR14d.pdf.

7. Piketty, *Capital in the Twenty-First Century*.
8. Tax Justice Network: https://en.wikipedia.org/wiki/Tax_Justice_Network.

APPENDIX

1. Ulijaszek et al., *The Cambridge Encyclopedia of Human Growth and Development*.

ALLEN LANE
an imprint of
PENGUIN BOOKS

Also Published

Justin Marozzi, *Islamic Empires: Fifteen Cities that Define a Civilization*

Bruce Hood, *Possessed: Why We Want More Than We Need*

Frank Close, *Trinity: The Treachery and Pursuit of the Most Dangerous Spy in History*

Janet L. Nelson, *King and Emperor: A New Life of Charlemagne*

Richard M. Eaton, *India in the Persianate Age: 1000-1765*

Philip Mansel, *King of the World: The Life of Louis XIV*

James Lovelock, *Novacene: The Coming Age of Hyperintelligence*

Mark B. Smith, *The Russia Anxiety: And How History Can Resolve It*

Stella Tillyard, *George IV: King in Waiting*

Donald Sassoon, *The Anxious Triumph: A Global History of Capitalism, 1860-1914*

Elliot Ackerman, *Places and Names: On War, Revolution and Returning*

Johny Pits, *Afropean: Notes from Black Europe*

Jonathan Aldred, *Licence to be Bad: How Economics Corrupted Us*

Walt Odets, *Out of the Shadows: Reimagining Gay Men's Lives*

Jonathan Rée, *Witcraft: The Invention of Philosophy in English*

Jared Diamond, *Upheaval: How Nations Cope with Crisis and Change*

Emma Dabiri, *Don't Touch My Hair*

Srecko Horvat, *Poetry from the Future: Why a Global Liberation Movement Is Our Civilisation's Last Chance*

Paul Mason, *Clear Bright Future: A Radical Defence of the Human Being*

Remo H. Largo, *The Right Life: Human Individuality and its role in our development, health and happiness*

Joseph Stiglitz, *People, Power and Profits: Progressive Capitalism for an Age of Discontent*

David Brooks, *The Second Mountain*

Roberto Calasso, *The Unnamable Present*

Lee Smolin, *Einstein's Unfinished Revolution: The Search for What Lies Beyond the Quantum*

Clare Carlisle, *Philosopher of the Heart: The Restless Life of Søren Kierkegaard*

Nicci Gerrard, *What Dementia Teaches Us About Love*

Edward O. Wilson, *Genesis: On the Deep Origin of Societies*

John Barton, *A History of the Bible: The Book and its Faiths*

Carolyn Forché, *What You Have Heard is True: A Memoir of Witness and Resistance*

Elizabeth-Jane Burnett, *The Grassling*

Kate Brown, *Manual for Survival: A Chernobyl Guide to the Future*

Roderick Beaton, *Greece: Biography of a Modern Nation*

Matt Parker, *Humble Pi: A Comedy of Maths Errors*

Ruchir Sharma, *Democracy on the Road*

David Wallace-Wells, *The Uninhabitable Earth: A Story of the Future*

Randolph M. Nesse, *Good Reasons for Bad Feelings: Insights from the Frontier of Evolutionary Psychiatry*

Anand Giridharadas, *Winners Take All: The Elite Charade of Changing the World*

Richard Bassett, *Last Days in Old Europe: Triste '79, Vienna '85, Prague '89*

Paul Davies, *The Demon in the Machine: How Hidden Webs of Information Are Finally Solving the Mystery of Life*

Toby Green, *A Fistful of Shells: West Africa from the Rise of the Slave Trade to the Age of Revolution*

Paul Dolan, *Happy Ever After: Escaping the Myth of The Perfect Life*

Sunil Amrith, *Unruly Waters: How Mountain Rivers and Monsoons Have Shaped South Asia's History*

Christopher Harding, *Japan Story: In Search of a Nation, 1850 to the Present*

Timothy Day, *I Saw Eternity the Other Night: King's College, Cambridge, and an English Singing Style*

Richard Abels, *Aethelred the Unready: The Failed King*

Eric Kaufmann, *Whiteshift: Populism, Immigration and the Future of White Majorities*

Alan Greenspan and Adrian Wooldridge, *Capitalism in America: A History*

Philip Hensher, *The Penguin Book of the Contemporary British Short Story*

Paul Collier, *The Future of Capitalism: Facing the New Anxieties*

Andrew Roberts, *Churchill: Walking With Destiny*

Tim Flannery, *Europe: A Natural History*

T. M. Devine, *The Scottish Clearances: A History of the Dispossessed, 1600-1900*

Robert Plomin, *Blueprint: How DNA Makes Us Who We Are*

Michael Lewis, *The Fifth Risk: Undoing Democracy*

Diarmaid MacCulloch, *Thomas Cromwell: A Life*

Ramachandra Guha, *Gandhi: 1914-1948*

Slavoj Žižek, *Like a Thief in Broad Daylight: Power in the Era of Post-Humanity*

Neil MacGregor, *Living with the Gods: On Beliefs and Peoples*

Peter Biskind, *The Sky is Falling: How Vampires, Zombies, Androids and Superheroes Made America Great for Extremism*

Robert Skidelsky, *Money and Government: A Challenge to Mainstream Economics*

Helen Parr, *Our Boys: The Story of a Paratrooper*

David Gilmour, *The British in India: Three Centuries of Ambition and Experience*

Jonathan Haidt and Greg Lukianoff, *The Coddling of the American Mind: How Good Intentions and Bad Ideas are Setting up a Generation for Failure*

Ian Kershaw, *Roller-Coaster: Europe, 1950-2017*

Adam Tooze, *Crashed: How a Decade of Financial Crises Changed the World*

Edmund King, *Henry I: The Father of His People*

Lilia M. Schwarcz and Heloisa M. Starling, *Brazil: A Biography*

Jesse Norman, *Adam Smith: What He Thought, and Why it Matters*

Philip Augur, *The Bank that Lived a Little: Barclays in the Age of the Very Free Market*

Christopher Andrew, *The Secret World: A History of Intelligence*

David Edgerton, *The Rise and Fall of the British Nation: A Twentieth-Century History*

Julian Jackson, *A Certain Idea of France: The Life of Charles de Gaulle*

Owen Hatherley, *Trans-Europe Express*

Richard Wilkinson and Kate Pickett, *The Inner Level: How More Equal Societies Reduce Stress, Restore Sanity and Improve Everyone's Wellbeing*

Paul Kildea, *Chopin's Piano: A Journey Through Romanticism*

Seymour M. Hersh, *Reporter: A Memoir*

Michael Pollan, *How to Change Your Mind: The New Science of Psychedelics*

David Christian, *Origin Story: A Big History of Everything*

Judea Pearl and Dana Mackenzie, *The Book of Why: The New Science of Cause and Effect*

David Graeber, *Bullshit Jobs: A Theory*

Serhii Plokhy, *Chernobyl: History of a Tragedy*

Michael McFaul, *From Cold War to Hot Peace: The Inside Story of Russia and America*

Paul Broks, *The Darker the Night, the Brighter the Stars: A Neuropsychologist's Odyssey*

Lawrence Wright, *God Save Texas: A Journey into the Future of America*

John Gray, *Seven Types of Atheism*

Carlo Rovelli, *The Order of Time*

Mariana Mazzucato, *The Value of Everything: Making and Taking in the Global Economy*

Richard Vinen, *The Long '68: Radical Protest and Its Enemies*

Kishore Mahbubani, *Has the West Lost It?: A Provocation*

John Lewis Gaddis, *On Grand Strategy*

Richard Overy, *The Birth of the RAF, 1918: The World's First Air Force*

Francis Pryor, *Paths to the Past: Encounters with Britain's Hidden Landscapes*

Helen Castor, *Elizabeth I: A Study in Insecurity*

Ken Robinson and Lou Aronica, *You, Your Child and School*

Leonard Mlodinow, *Elastic: Flexible Thinking in a Constantly Changing World*

Nick Chater, *The Mind is Flat: The Illusion of Mental Depth and The Improvised Mind*

Michio Kaku, *The Future of Humanity: Terraforming Mars, Interstellar Travel, Immortality, and Our Destiny Beyond*

Thomas Asbridge, *Richard I: The Crusader King*

Richard Sennett, *Building and Dwelling: Ethics for the City*

Nassim Nicholas Taleb, *Skin in the Game: Hidden Asymmetries in Daily Life*

Steven Pinker, *Enlightenment Now: The Case for Reason, Science, Humanism and Progress*

Steve Coll, *Directorate S: The C.I.A. and America's Secret Wars in Afghanistan, 2001 - 2006*

Jordan B. Peterson, *12 Rules for Life: An Antidote to Chaos*

Bruno Maçães, *The Dawn of Eurasia: On the Trail of the New World Order*

Brock Bastian, *The Other Side of Happiness: Embracing a More Fearless Approach to Living*

Ryan Lavelle, *Cnut: The North Sea King*

Tim Blanning, *George I: The Lucky King*

Thomas Cogswell, *James I: The Phoenix King*

Pete Souza, *Obama, An Intimate Portrait: The Historic Presidency in Photographs*

Robert Dallek, *Franklin D. Roosevelt: A Political Life*

Norman Davies, *Beneath Another Sky: A Global Journey into History*

Ian Black, *Enemies and Neighbours: Arabs and Jews in Palestine and Israel, 1917-2017*

Martin Goodman, *A History of Judaism*

Shami Chakrabarti, *Of Women: In the 21st Century*

Stephen Kotkin, *Stalin, Vol. II: Waiting for Hitler, 1928-1941*

Lindsey Fitzharris, *The Butchering Art: Joseph Lister's Quest to Transform the Grisly World of Victorian Medicine*

Serhii Plokhy, *Lost Kingdom: A History of Russian Nationalism from Ivan the Great to Vladimir Putin*

Mark Mazower, *What You Did Not Tell: A Russian Past and the Journey Home*

Lawrence Freedman, *The Future of War: A History*

Niall Ferguson, *The Square and the Tower: Networks, Hierarchies and the Struggle for Global Power*

Matthew Walker, *Why We Sleep: The New Science of Sleep and Dreams*

Edward O. Wilson, *The Origins of Creativity*

John Bradshaw, *The Animals Among Us: The New Science of Anthropology*